CONSTRUCTIONS

Cognitive Theory of Language and Culture

A series edited by Gilles Fauconnier,

George Lakoff, and Eve Sweetser

CONSTRUCTIONS

A Construction Grammar

Approach to

Argument Structure

Adele E. Goldberg

The University of Chicago Press

Chicago and London

ADELE E. GOLDBERG is assistant professor of linguistics at the University of California, San Diego.

The University of Chicago Press, Chicago 60637
The University of Chicago Press, Ltd., London

© 1995 by The University of Chicago
All rights reserved. Published 1995
Printed in the United States of America

04 03 02 01 00 99 98 97 96 95   5 4 3 2 1

ISBN (cloth): 0-226-30085-4
ISBN (paper): 0-226-30086-2

Library of Congress Cataloging-in-Publication Data

Goldberg, Adele E.
    Constructions : a construction grammar approach to argument structure /
Adele E. Goldberg.
        p.   cm. — (Cognitive theory of language and culture)
    Originally presented as the author's thesis (Ph.D.)—University of California,
1992.
    Includes bibliographical references (p.) and index.
    ISBN 0-226-30085-4 (cloth). — ISBN 0-226-30086-2 (pbk.)
    1. Grammar, Comparative and general—Syntax.   2. Semantics.
3. Generative grammar.   I. Title.   II. Series.
P291.G65   1995
415—dc20   94-20705
                                                                              CIP

∞ The paper used in this publication meets the minimum requirements of the American National Standard for Information Sciences—Permanence of Paper for Printed Library Materials, ANSI Z39.48-1984.

This book is printed on acid-free paper.

To Ali

Contents

Acknowledgments

This book grew out of my Ph.D. thesis (Goldberg 1992b), which was completed at the University of California, Berkeley. An enormous debt is owed to my advisor, George Lakoff, for his wisdom, enthusiasm, and encouragement, his ever-ready example and counterexample, and for sharing his time and his deep insights with incredible generosity.

I'd like to thank Charles Fillmore for instilling in me a deep respect for the complexities of the data, and for sharing his wisdom. His enduring insights have profoundly influenced this work in innumerable ways. I'm also grateful for his spearheading the development of the theory of Construction Grammar, on which the present work is based.

Work in Construction Grammar includes, for example, Fillmore, Kay and O'Connor's analysis of the *let alone* and *the more, the merrier* constructions (1988), Brugman's analysis of *have* constructions (1988), Kay's work on *even* (1990), the "What, me worry?" construction of Lambrecht (1990), and Sweetser's analysis of modal verbs (1990). Construction Grammar is also developed in Fillmore (1985b, 1987, 1988, 1990), Fillmore & Kay (1993), Filip (1993), Jurafsky (1992), Koenig (1993), and Michaelis (1993). The present work owes its greatest debts to Lakoff's in-depth study of *there* constructions (1984) and to Fillmore (1987), who suggested that the meaning of an expression is arrived at by the superimposition of the meanings of open class words with the meanings of grammatical elements.

I'm grateful to Dan Slobin for his encouragement and guidance, and for providing a reality check on the plausibility of psychological claims. In the final stages of writing my dissertation, I was fortunate enough to work closely with Annie Zaenen. I am immensely grateful for her advice, her many leads to relevant literature, and for our many interesting and helpful discussions, which have deeply influenced my work.

Other members of the Berkeley faculty contributed in countless ways to my education. Eve Sweetser tirelessly read and offered valuable comments on many papers; Paul Kay provided much helpful input, and was consistently willing to lend an ear and a critical eye; Robert Wilensky offered many helpful discussions and some wonderful data. Len Talmy was always willing to discuss all manner of ideas. Visitors Don Forman, Knud Lambrecht, Minoko Nakau,

Frederika Van der Leek, and Robert Van Valin offered different perspectives and very helpful discussions.

I'd like to offer personal thanks to Claudia Brugman, Michele Emanatian, Hana Filip, Jean-Pierre Koenig, and Laura Michaelis for support of every kind, including countless enlightening discussions on topics related to almost every aspect of this monograph. I'm also grateful to Jess Gropen, Beth Levin, Steve Pinker, and Ray Jackendoff, for their own inspirational work and for their helpful feedback and discussion.

During the writing and rewriting of this manuscript, I was able to spend a good deal of time at Berkeley, Xerox Palo Alto Research Center, Stanford, and the University of California, San Diego, so there are many people to thank for very helpful suggestions and discussions, including Farrell Ackerman, Joan Bresnan, Tony Davis, Jane Espenson, Gilles Fauconnier, Joe Grady, Marti Hearst, Kyoko Hirose, Rolf Johnson, Dan Jurafsky, Suzanne Kemmer, Yuki Kuroda, Ron Langacker, Maarten Lemmens, John Moore, Terry Regier, Hadar Shem-Tov, Eve Clark, Cleo Condoravdi, Mark Gawron, Jess Gropen, Geoff Nunberg, Ivan Sag, Tom Wasow, Ali Yazdani, and Sandro Zucchi. Several UCSD students carefully read the manuscript and made very helpful suggestions, particularly Kathleen Ahrens, Michael Israel, and Bill Morris. For help preparing the manuscript I would like to thank Kathleen Ahrens, Bill Byrne, and Nitya Sethuraman. Finally, for editorial assistance I thank Geoff Huck and Karen Peterson, and for the most careful, well-informed copy-editing I could have imagined, I thank Christine Bartels.

Excerpts of this book first appeared, in different form, as articles or book chapters. I thank the publishers for permission to include revised material from: "The Inherent Semantics of Argument Structure: The Case of the English Ditransitive Construction," *Cognitive Linguistics* 3(1):37–74, 1992; "A Semantic Account of Resultatives," *Linguistic Analysis* 21:66–96, 1991; "It Can't Go Down the Chimney Up: Paths and the English Resultative," *BLS* 17; "Making One's Way Through the Data," in A. Alsina, J. Bresnan, and P. Sells (eds.), *Complex Predicates,* CSLI Publications, forthcoming; "Another Look at Some Learnability Paradoxes," in *Proceedings of the 25th Annual Stanford Child Language Research Forum,* CSLI Publications.

For providing financial support, comfortable offices, and stimulating environments, I'd like to thank the Sloan Foundation, who funded the Cognitive Science Institute at Berkeley, the International Computer Science Institute (ICSI), the Center for the Study of Language and Information (CSLI), and Xerox PARC.

A crucial debt is owed to my family: my mom, Ann Goldberg, for being a voice of reason on topics related to this monograph and on all others; my siblings, Ken Goldberg and Elena Goldberg Man, and my grandparents, Harry and Birdie Goldberg and Rose Wallach, for their consistent loving support, and just for being themselves. I am also deeply grateful to the memory of my father, Melvin Goldberg, for his unparalleled courage, curiosity, and compassion.

Finally, I am immensely grateful to Ali Yazdani, for always being there, even though there have been many miles between us. This book is dedicated to him.

1 Introduction

1.1 THE CONCEPT OF CONSTRUCTIONS

What is it children learn when they learn to speak a language? What is the nature of verb meaning and what is its relation to sentential meaning? How and to what extent are novel utterances based on previously learned utterances?

These questions are addressed here through a study of basic sentence types—the "simple sentences" of traditional grammarians. A central thesis of this work is that basic sentences of English are instances of *constructions*—form–meaning correspondences that exist independently of particular verbs. That is, it is argued that constructions themselves carry meaning, independently of the words in the sentence.

The notion *construction* has a time-honored place in linguistics. Traditional grammarians have inevitably found it useful to refer to properties of particular constructions. The existence of constructions in the grammar was taken to be a self-evident fact that required little comment. In the early stages of transformational grammar (Chomsky 1957, 1965), constructions retained their central role, construction-specific rules and constraints being the norm. In the past two decades, however, the pretheoretical notion of construction has come under attack. Syntactic constructions have been claimed to be epiphenomenal, arising solely from the interaction of general principles (Chomsky 1981, 1992); the rejection of constructions in favor of such general principles is often assumed now to be the only way to capture generalizations across patterns.

At the same time, the rising tide of interest in semantic and pragmatic properties has led to a renewed focus on the idiosyncratic properties of particular sentence patterns (cf. Levin 1993, for example). In order to reconcile the theoretical desire for construction-independent principles with the empirical necessity of recognizing pattern-specific properties, all such idiosyncratic properties have been attributed to individual lexical items, lexical entries being the last refuge of the idiosyncratic.

There is no question that a large amount of information is contributed by individual lexical items (cf. chapters 2 and 5). However, in this work it is argued that an entirely lexically-based, or bottom-up, approach fails to account for the full range of English data. Particular semantic structures together with their associated formal expression must be recognized as constructions independent of the lexical items which instantiate them.

This monograph thus represents an effort to bring constructions back to their rightful place on center stage by arguing that they should be recognized as theoretical entities. Single-clause patterns hold a special interest because these cases clearly lie at the heart of any theory of grammar. If it can be shown that constructions are essential to a description of the domain of simple clauses, then it must be recognized that constructions are crucial to the description of language. Chapters 3 and 4 argue that empirical generalizations across constructions can in fact naturally be captured within a construction-based framework.

Another goal of this monograph is to explicate the semantics associated with particular clausal patterns. The semantic properties to be discussed must be accounted for by any framework, regardless of where the semantics is encoded or what one's assumptions about the lexicon and syntax are.

It has long been recognized that differences in complement configuration are often associated with differences in meaning. For example, the ditransitive requires that its goal argument be animate, while the same is not true of paraphrases with *to:*

(1)   a.   I brought Pat a glass of water. (ditransitive)
       b.   I brought a glass of water to Pat.
(2)   a.   *I brought the table a glass of water. (ditransitive)
       b.   I brought a glass of water to the table. (Partee 1965:60)

Fillmore (1968, fn. 49) noted that sentences such as the following differ in meaning:

(3)   a.   Bees are swarming in the garden.
       b.   The garden is swarming with bees.

(3b) suggests that the whole garden is full of bees, whereas (3a) could involve bees in only a part of the garden.

Anderson (1971) observed that the following sentences also differ in meaning:

(4)   a.   I loaded the hay onto the truck.
       b.   I loaded the truck with the hay.

While (4b) implies that the truck is entirely filled with hay (or at least relevantly affected), no such implication exists in (4a).

Works by Green, Oehrle, Bolinger, Borkin, and Wierzbicka and by Interpretive Semanticists such as Chomsky, Partee, and Jackendoff have drawn attention to systematic differences in meaning between sentences with the same lexical items in slightly different constructions.[1] Borkin (1974), for example, provides the following contrast:

(5) a. When I looked in the files, I found that she was Mexican.
   b. ?When I looked in the files I found her to be Mexican.
   c. *When I looked in the files I found her Mexican.

Borkin argues that the pattern in (5c) is only possible with verbs of proposition when the proposition expressed is considered to be a matter of judgment, as opposed to a matter of fact. The pattern in (5b) prefers but does not require the proposition to express judgments, and the full clausal form with *that*-complementizer in (5a) freely allows matters of judgment or fact.

Wierzbicka (1988) contrasts (6a) and (6b):

(6) a. I am afraid to cross the road.
   b. I am afraid of crossing the road.

Only in (6a) is the speaker presumed to have some intention of crossing the road. This difference in interpretation is argued to account for why (7a) is infelicitous unless the falling is interpreted as somehow volitionally intended:[2]

(7) a. #I am afraid to fall down.
   b. I am afraid of falling down.

Similar observations of subtle differences in meaning led Bolinger to conclude: "A difference in syntactic form always spells a difference in meaning" (1968:127). The same hypothesis—which we may term the Principle of No Synonymy of Grammatical Forms—has been formulated by Givón (1985), Kirsner (1985), Langacker (1985), Clark (1987), and Wierzbicka (1988). It will be adopted here as a working hypothesis.[3]

◆

In this monograph, I explore the idea that *argument structure constructions* are a special subclass of constructions that provides the basic means of clausal expression in a language.[4] Examples of English argument structure constructions to be discussed here include the following:

1. Ditransitive	X CAUSES Y to RECEIVE Z	Subj V Obj Obj$_2$
		Pat faxed Bill the letter.
2. Caused Motion	X CAUSES Y to MOVE Z	Sub V Obj Obl
		Pat sneezed the napkin off the table.
3. Resultative	X CAUSES Y to BECOME Z	Subj V Obj Xcomp
		She kissed him unconscious.
4. Intrans. Motion	X MOVES Y	Subj V Obl
		The fly buzzed into the room.

5. Conative   X DIRECTS ACTION at Y   Subj V Obl$_{at}$
                                                Sam kicked at Bill.

On a constructional approach to argument structure, systematic differences in meaning between the same verb in different constructions are attributed directly to the particular constructions. We will see that if we consider various constructions on their own terms, interesting generalizations and subtle semantic constraints emerge. Several constructions can be shown to be associated with a family of distinct but related senses, much like the polysemy recognized in lexical items. Moreover, these constructions themselves are shown to be interrelated.

The analysis I am going to propose draws on research in Construction Grammar (cf. Fillmore 1985b, 1987, 1988, 1990; Fillmore & Kay 1993; Lakoff 1987; Fillmore, Kay & O'Connor 1988; Brugman 1988; Kay 1990; Lambrecht 1990, 1994; Goldberg 1991a, 1992a; Michaelis 1993; Koenig 1993; Filip 1993). According to Construction Grammar, a distinct construction is defined to exist if one or more of its properties are not strictly predictable from knowledge of other constructions existing in the grammar:[5]

> C is a CONSTRUCTION iff$_{def}$ C is a form–meaning pair $<F_i, S_i>$ such that some aspect of F_i or some aspect of S_i is not strictly predictable from C's component parts or from other previously established constructions.

Constructions are taken to be the basic units of language. Phrasal patterns are considered constructions if something about their form or meaning is not strictly predictable from the properties of their component parts or from other constructions.[6] That is, a construction is posited in the grammar if it can be shown that its meaning and/or its form is not compositionally derived from other constructions existing in the language (cf. section 1.2). In addition, expanding the pretheoretical notion of construction somewhat, morphemes are clear instances of constructions in that they are pairings of meaning and form that are not predictable from anything else (Saussure 1916).[7] It is a consequence of this definition that the lexicon is not neatly differentiated from the rest of grammar.

Constructions can be understood to correspond to the "listemes" of DiSciullo and Williams (1987)—that is, the entities of grammar that must be listed. However, our view of the collection of listemes is radically different from theirs. They state categorically: "If conceived of as the set of listemes, the lexicon is incredibly boring by its very nature. It contains objects of no single specifiable type (words, VPs, morphemes, perhaps intonational patterns, and so on), and those objects that it does contain are there because they fail to conform to interesting laws. The lexicon is like a prison—it contains only the lawless, and the only thing that its inmates have in common is lawlessness" (p. 3). This view of the lexicon, or what might be better termed the *constructicon,* following Jurafsky (1992), is rejected in the present work. The collection of constructions is not assumed to consist of an unstructured set of independent entities, but instead it is taken to constitute a highly structured lattice of interrelated information. The relations between constructions are discussed in chapters 3 and 4.

A basic axiom that is adopted is: knowledge of language is knowledge. Many of the findings of the following chapters are thus expected, particularly that linguistic constructions display prototype structure and form networks of associations. Hierarchies of inheritance and semantic networks, long found useful for organizing other sorts of knowledge, are adopted for explicating our linguistic knowledge (cf. Quillian 1968; Bobrow & Winograd 1977; Fahlman 1979; Wilensky 1986; Norvig & Lakoff 1987; Jurafsky 1992).

On the basis of research on language acquisition by Clark (1978), Slobin (1985), and Bowerman (1989), together with the findings presented here, it is hypothesized that

> Simple clause constructions are associated directly with semantic structures which reflect scenes basic to human experience.[8]

In particular, constructions involving basic argument structure are shown to be associated with dynamic scenes: experientially grounded gestalts, such as that of someone volitionally transferring something to someone else, someone causing something to move or change state, someone experiencing something, something moving, and so on. It is proposed that the basic clause types of a language form an interrelated network, with semantic structures paired with particular forms in as general a way as possible.

◆

This book is structured as follows. The rest of this chapter presents arguments for adopting a constructional approach to argument structure. Chapter 2 analyzes the nature of verb meaning, the nature of constructional meaning, and the relation between the two. Chapter 3 suggests an account of how to capture relations among constructions and generalizations across constructions; an inheritance hierarchy of constructions is posited, and the inheritance links themselves are treated as objects in the system. In chapter 4, the idea of a monostratal theory is defended, and the way linking generalizations are to be captured within a constructional approach is discussed. Chapter 5 presents an account of the partial productivity of constructions; this work adapts insights from Pinker (1989) to a system without lexical rules.

Chapters 6–9 involve more specific analyses of several English constructions: the ditransitive construction (e.g., *Chris faxed her the news*), the "caused-motion" construction (e.g., *Sally sneezed the napkin off the table*), the resultative construction (e.g., *Sam talked himself hoarse*), and the *way* construction (e.g., *Bob elbowed his way through the crowd*). Specific arguments for the existence of each of these constructions are given in those chapters.

1.2   A Brief Introduction to Construction Grammar

The basic tenet of Construction Grammar as developed in Fillmore & Kay 1993, Fillmore, Kay & O'Connor 1988, Lakoff 1987, Brugman 1988, Lambrecht 1994, is that traditional constructions—i.e., form–meaning correspondences—are the basic units of language.

Theorists working within this theory share an interest in characterizing the *entire* class of structures that make up language, not only the structures that are defined to be part of "core grammar." This interest stems from the belief that fundamental insights can be gained from considering such non-core cases, in that the theoretical machinery that accounts for non-core cases can be used to account for core cases. In addition, much of actual corpus data involves such non-core cases. Construction Grammarians also share an interest in accounting for the conditions under which a given construction can be used felicitously, since this is taken to be part of speakers' competence or knowledge of language; from this interest stems the conviction that subtle semantic and pragmatic factors are crucial to understanding the constraints on grammatical constructions.

These tenets, which in many respects hearken back to Generative Semantics (e.g. Lakoff 1965, 1970a,b, 1971, 1972, 1976; Lakoff & Ross 1976; Langacker 1969; Postal 1971; Dowty 1972; Keenan 1972; McCawley 1973, 1976) are also shared by the theory of Cognitive Grammar (Langacker 1987a, 1991), the framework implicit in much of Wierzbicka's work (e.g., Wierzbicka 1988), and by many functionalist approaches to grammar (e.g., Bolinger 1968; DeLancey 1991; Givón 1979a,b; Haiman 1985a; Foley & Van Valin 1984). Work in Generalized Phrase Structure Grammar (GPSG) and in Head-Driven Phrase Structure Grammar (HPSG) (Gazdar et al. 1985; Pollard & Sag 1987, 1994) also emphasizes the central role of the *sign* in grammar. In many ways, aspects of the proposals made here are also compatible with recent work by Levin (1985), Levin & Rapoport (1988), Pinker (1989) and Jackendoff (1990a). Some similarities and differences are discussed below.

Owing in part to the fact that Construction Grammar has grown largely out of work on frame semantics (Fillmore 1975, 1977b, 1982, 1985a) and an experientially based approach to language (Lakoff 1977, 1987), the approach to semantics that is adopted by the theory is one that crucially recognizes the importance of speaker-centered "construals" of situations in the sense of Langacker (1987a, 1991). This approach to semantics is discussed in chapter 2.

◆

In Construction Grammar, no strict division is assumed between the lexicon and syntax. Lexical constructions and syntactic constructions differ in internal complexity, and also in the extent to which phonological form is specified, but both lexical and syntactic constructions are essentially the same type of declaratively represented data structure: both pair form with meaning. It is not the case, however, that in rejecting a strict division, Construction Grammar denies the existence of any distinctly morphological or syntactic constraints (or constructions). Rather, it is claimed that there are basic commonalities between the two types of constructions, and moreover, that there are cases, such as verb–particle combinations, that blur the boundary.

Another notion rejected by Construction Grammar is that of a strict division between semantics and pragmatics. Information about focused constituents, topicality, and register is represented in constructions alongside semantic information.

Construction Grammar is generative in the sense that it tries to account for the infinite number of expressions that are allowed by the grammar while attempting to account for the fact that an infinite number of other expressions are ruled out or disallowed. Construction Grammar is not transformational. No underlying syntactic or semantic forms are posited. Instead, Construction Grammar is a monostratal theory of grammar like many other current theories, including Lexical Functional Grammar (LFG) (Bresnan 1982), Role and Reference Grammar (Foley & Van Valin 1984), GPSG (Gazdar et al. 1985), HPSG (Pollard & Sag 1987, 1994), and Cognitive Grammar (Langacker 1987a, 1991). The rationale for this and some consequences are discussed in chapter 4.

It is perhaps easiest to explore the constructional approach by first contrasting it with the relevantly similar proposal described in the following section.

1.3 An Alternative Account: Lexicosemantic Rules

The recognition of subtle semantic differences between related syntactic (subcategorization) frames has been growing, and there has also been increasing focus on the fact that there appears to be a strong correlation between the meanings of verbs and the syntactic frames they can occur in, leading many researchers to speculate that in any given language the syntactic subcategorization frames of a verb may be uniquely predictable from the verb's lexical semantics (e.g., Levin 1985; Chomsky 1986; Carter 1988; Levin & Rapoport 1988; Rappaport & Levin 1988; Pinker 1989; Gropen et al. 1989).

The following factors have led these theorists to postulate lexical rules which are designed to operate on the semantic structures of lexical items: (1) overt complement structure appears to be predictable by general linking rules that map semantic structure onto syntactic form, and (2) the same verb stem often occurs with more than one complement configuration.

For example, Pinker (1989) proposes that the prepositional/ditransitive alternation (the "dative" alternation) results from a semantic rule rather than being the product of a syntactic transformation. Specifically, he suggests that productive use of the ditransitive syntax is the result of a lexicosemantic rule which takes as input a verb with the semantics 'X CAUSES Y to GO TO Z' and produces the semantic structure 'X CAUSES Z to HAVE Y'. The double object syntax, he argues, is then predictable from near-universal linking rules mapping the arguments of a verb with the meaning 'X CAUSES Z to HAVE Y' into the ditransitive form. In this way, Pinker argues that the dative rule produces a "conceptual gestalt shift,"—that it is, in effect, a semantic operation on lexical structure (cf. also Gropen et al. 1989).

The general approach can be outlined as follows:

1a. The syntactic complement configuration of a clause is taken to be uniquely predictable from the semantic representation of the matrix verb. The mapping from semantic representations to particular complement configurations is performed via universal, or near-universal, linking rules.

1b. Different syntactic complement configurations therefore reflect differences in the semantic representations of the main verb.

2. Different semantic representations of a particular verb stem, i.e., different verb senses, are related by generative lexical rules which take as input a verb with a particular semantics and yield as output a verb with a different semantics.

3. Differences in semantics are not necessarily truth-functional differences, but may represent a different construal of the situation being described; that is, the relevant semantics is speaker-based.

These principles are detailed most explicitly in Pinker 1989, but are also shared by Levin 1985, Levin & Rapoport 1988, and Gropen et al. 1989.

By postulating rules that operate on semantic structure, as opposed to rules or transformations that are purely or primarily syntactic, these theories manage to incorporate important insights. As was discussed above, different constructions are typically, possibly always, accompanied by slightly different semantic interpretations; these semantic differences are respected as soon as the forms are learned (Bowerman 1982; Gropen et al. 1989). By postulating semantics-changing rules, as opposed to syntactic rules with additional semantic constraints, such theories capture the insight that changes in complement configurations are crucially semantic. Regularities in the syntax are captured by linking rules mapping the semantic structure to surface form.

To a large degree, as will become apparent below, the lexical rule approach is directly comparable to the approach being proposed here. They share the emphasis on semantic differences among different complement configurations. The strongest differences between the two approaches stem from the increased focus of the present approach on the nature of the relation between verb and construction (the lexical rule approach represents this relation only implicitly in the statement of the rule itself). By recognizing constructions and verbs to be interrelated but independent, the nature of constructional meaning, the principles that relate verb and construction, and the relations among constructions are brought to the foreground. These topics are the focus of much of the present work. In addition, on the present approach it is not necessary to posit an additional verb sense for each new syntactic configuration in which the verb appears. Several general reasons to prefer the constructional approach to the lexical rule approach just described are detailed in the following section. Specific arguments for the existence of each construction analyzed in chapters 6–9 are provided in those chapters.

1.4 ADVANTAGES OF THE CONSTRUCTION ACCOUNT

1.4.1 Implausible Verb Senses Are Avoided

The constructional approach avoids the problem of positing implausible verb senses to account for examples such as the following:

(8) He sneezed the napkin off the table.
(9) She baked him a cake.
(10) Dan talked himself blue in the face.

In none of these cases does the verb intuitively require the direct object complement. To account for (8), for example, a lexicosemantic theory would have to say that *sneeze,* a parade example of an intransitive verb, actually has a three-argument sense, 'X CAUSES Y to MOVE Z by sneezing'. To account for (9), such a theory would need to claim that there exists a special sense of *bake* that has three arguments: an agent, a theme, and an intended recipient. This in effect argues that *bake* has a sense which involves something like 'X INTENDS to CAUSE Y to HAVE Z'. To account for (10), the theory would need to postulate a special sense of *talk,* 'X CAUSES Y to BECOME Z by talking'.

If additional senses were involved, then it would follow that each of these verbs is ambiguous between its basic sense and its sense in the syntactic pattern above. Therefore we would expect that there would be some language that differentiates between the two senses by having two independent (unrelated) verb stems. For example, alongside the equivalent of the English word *sneeze* we might expect to find another stem—say, *moop*—that meant 'X CAUSES Y to MOVE Z by sneezing'. However, to my knowledge there is no language that has distinct verb stems for any of the meanings represented by examples (8–10).

On a constructional approach, we can understand aspects of the final interpretation involving caused motion, intended transfer, or caused result to be contributed by the respective constructions. That is, we can understand skeletal constructions to be capable of contributing arguments. For example, we can define the ditransitive construction to be associated directly with agent, patient, and recipient roles, and then associate the class of verbs of creation with the ditransitive construction. We do not need to stipulate a specific sense of *bake* unique to this construction. In general, we can understand the direct objects found in the above examples to be licensed not directly as arguments of the verbs but by the particular constructions. This idea is discussed in more detail in chapter 2.

Other examples where it is implausible to attribute the complement configuration and the resulting interpretation directly to the main verb include the following:

(11) "Despite the President's efforts to *cajole* or *frighten* his nine million subjects into line . . ." (*New York Times,* 29 May 1993)

(12) "My father *frowned* away the compliment and the insult." (Stephen McCauley, *Easy Way Out,* 1993)

(13) "Sharon was exactly the sort of person who'd *intimidate* him into a panic." (Stephen McCauley, *Easy Way Out,* 1993)

(14) "I cannot inhabit his mind nor even *imagine* my way through the dark labyrinth of its distortion." (Oxford University Press corpus)

(15) Pauline *smiled* her thanks. (Levin & Rapoport 1988)

(16) The truck *rumbled* down the street. (Levin & Rappaport Hovav 1990b)

The suggestion being made here is to account for these cases, in which the whole is not built up from the lexical items in a straightforward way, by postulating a construction that is itself associated with meaning.

1.4.2   Circularity Is Avoided

Another important advantage of the construction-based approach is that it avoids a certain circularity of analysis resulting from the widespread claim in current linguistic theories that syntax is a projection of lexical requirements.

This claim is explicit in the Projection Principle of Government and Binding Theory (GB) (Chomsky 1981), the Bijection Principle of Lexical Functional Grammar (Bresnan 1982), and in all current accounts which attempt to predict overt syntax from semantic roles or theta role arrays. In all of these frameworks, it is the verb which is taken to be of central importance. That is, it is assumed that the verb determines how many and which kinds of complements will co-occur with it. In this way, the verb is analogized to the predicate of formal logic, which has an inherent number of distinct arguments. The verb is taken to be an n-place relation "waiting" for the exactly correct type and number of arguments. But note, now, that an ordinary verb such as *kick* can appear with at least eight distinct argument structures:

1. Pat kicked the wall.
2. Pat kicked Bob black and blue.
3. Pat kicked the football into the stadium.
4. Pat kicked at the football.
5. Pat kicked his foot against the chair.
6. Pat kicked Bob the football.
7. The horse kicks.
8. Pat kicked his way out of the operating room.

Theories which assume that the verb directly determines particular complement configurations are forced to claim that *kick* is a binary relation with agent and patient arguments and therefore occurs with transitive syntax, except in *Pat kicked Bob the football,* in which it is a ternary relation with agent, recipient, and patient arguments and therefore occurs in the ditransitive construction, and in *Pat kicked the football into the stadium,* where *kick* is again ternary, but now with agent, theme, and goal arguments, and must "therefore" occur with the direct object and prepositional complements; and so on. Thus both the evidence for the claim that *kick* has a particular n-argument sense *and* the explanation for *kick* having the corresponding complement configuration come from the fact that *kick* can occur overtly with a particular n-complement construction. That is, it is claimed that *kick* has an n-argument sense on the basis of the fact that *kick* occurs with n complements; it is simultaneously argued that *kick* occurs with n complements because it has an n-argument sense. This is where the circularity arises.

A constructional approach to argument structure allows us to avoid the circularity of arguing that a verb is an n-ary predicate and "therefore" has n complements when and only when it has n complements. Instead, the ternary relation, for example, is directly associated with the skeletal ditransitive construction. The verb, on the other hand, is associated with one or a few basic senses which must be *integrated* into the meaning of the construction. Under what conditions this is possible is the subject of the following chapter. Instead of positing a new sense every time a new syntactic configuration is encountered and then using that sense to explain the existence of the syntactic configuration, a constructional approach requires that the issue of the interaction between verb meaning and constructional meaning be addressed.

1.4.3   Semantic Parsimony

Levin (1985) suggests that evidence for different verb senses does exist. For example, she argues that "there is evidence that when the verb *slide* is found in the double object construction, . . . its sense is not the purely physical transfer sense of *slide* but rather a transfer of possession sense" (p. 35). She cites the fact that "the goal argument of a change of possession verb must denote an entity capable of ownership, but the goal argument of a change of location verb need not," as illustrated by her examples (17a, b).

(17)   a.   She slid Susan/*the door the present.
       b.   She slid the present to Susan/to the door.

Thus two distinct senses of *slide* would be posited to account for the contrast in (17). One sense of *slide,* 'slide$_1$', would constrain its goal to be animate, while the other, 'slide$_2$', would have no such constraint. The two different syntactic realizations are claimed to follow from universal or near-universal linking patterns mapping semantic argument structures to overt complement configurations. The linking rules would be sensitive to the fact that 'slide$_1$' requires its goal to be animate, as follows:

$$\text{'slide}_1\text{': } <\text{agt, pat, goal}_{animate}> \qquad\qquad \text{'slide}_2\text{': } <\text{agt, pat, goal}>$$
$$\downarrow \qquad\qquad \text{Linking Rules} \qquad\qquad \downarrow$$
$$\text{She slid Susan the present.} \qquad\qquad \text{She slid the present to Susan.}$$

However, general linking rules do not insure that 'slide$_1$' will only occur with the ditransitive construction, as is desired. Verbs which uncontroversially *lexically* constrain their goals to be animate—such as *give* or *hand*—can be used with both syntactic patterns:

$$\text{hand: } <\text{agt, pat, goal}_{animate}> \; (\approx \text{'slide}_1\text{')}$$
$$\swarrow \qquad\qquad\qquad \searrow$$
$$\text{Joe handed his mother a letter.} \qquad \text{Joe handed a letter to his mother.}$$

That is, we would need to stipulate that 'slide$_1$' may only occur with the ditransitive construction. Instead of positing both an additional sense of *slide* and a stipulation that this sense can only occur in the ditransitive construction, we can attribute the constraint that the goal must be animate directly to the construction.

Still, it might be argued that 'slide₁' is not actually constrained to appear ditransitively, and that it is this sense which (just like *give* and *hand*) appears in expressions such as (18):

(18)   She slid the present to Susan.

(The reason we might assume that (18) involves an unconstrained sense of *slide* is that *She slid the present to the door* is also acceptable.) This does not alleviate the problem, however; we still need to insure that the ditransitive construction can only occur with 'slide₁'. That is, instead of needing to stipulate that 'slide₁' can only appear ditransitively, we would now need to posit a constraint on the construction that permits it to only occur with verbs which constrain their goals to be animate. But with this constraint in place, there is no need to posit an additional verb sense.

More generally, I concur with Levin that the semantics of (and constraints on) the full expressions are different whenever a verb occurs in a different construction. But these differences need not be attributed to different verb senses; they are more parsimoniously attributed to the constructions themselves.

1.4.4   Compositionality Is Preserved

A construction is posited in the grammar if and only if something about its form, meaning, or use is not strictly predictable from other aspects of the grammar, including previously established constructions. In order to understand this principle, we must first consider the notion of *compositionality*. Frege is generally acknowledged to have originally formulated the idea that semantics need be compositional: the meaning of every expression in a language must be a function of the meanings of its immediate constituents and the syntactic rule used to combine them.

Montague stated the analogous condition that there must be a homomorphism from syntax to semantics; that is, there must be a structure-preserving mapping from syntax to semantics. Letting σ be a function from syntax to semantics, '$+_{\text{syn-comp}}$' a rule of syntactic composition, and '$+_{\text{sem-comp}}$' a rule of semantic composition, the following is claimed hold:

(19) $\sigma(x +_{\text{syn-comp}} y) = \sigma(x) +_{\text{sem-comp}} \sigma(y)$

The meaning of the expression is therefore taken to result from applying to the meanings of the immediate constituents a semantic operation which directly corresponds to the relevant syntactic operation.

Dowty (1979) observes that the claim is intended to imply that the relation between syntactic expression and semantic representation is straightforward and direct. That is, '$+_{\text{syn-comp}}$', or syntactic composition, must be straightforwardly related to '$+_{\text{sem-comp}}$', or semantic composition. The same principle, that the semantic rules of combination must directly reflect the syntactic rules of combination, is expressed by Gazdar et al. (1985), also working within the Montague Grammar tradition: "We assume that there exists a universal mapping from syntactic rules to semantic translations . . . . We claim that the semantic type assigned to any lexical item introduced in a rule . . . and the syntactic form of the rule itself are sufficient to fully determine . . . the form of the semantic translation rule" (1985:8–9).

Because the rules of combination are so widely regarded as transparent, it is easy to overlook the fact that there are any substantive rules at all. For example, one researcher states: "In a strictly compositional language, all analytic content comes from the lexicon, and no semantic rules . . . are needed to account . . . [for the mechanism of] adding meaning to the sentence which is not directly contributed by some lexeme of the sentence." [9]

Even Jackendoff, who in fact does recognize nonlexical meaning (cf. section 10.1.1), states in the introduction to his 1990 monograph *Semantic Structures:* "It is widely assumed, and I will take for granted, that the basic units out of which a sentential concept is constructed are the concepts expressed by the words in the sentence, that is, *lexical* concepts" (Jackendoff 1990a:9). The transparent rule of composition for verbs that is typically assumed goes back to Frege (1879): the meaning of a verb is a predicate with a fixed arity n that takes n arguments and yields a proposition. In this way, the verb is taken to be the semantic *head* of the sentence, the element which determines the basic semantic structure of the clause.

This same idea is implemented in recent *unification-based* grammars (cf. Shieber et al. 1984; Shieber 1986), for example, LFG, GPSG, and HPSG, which make explicit the critical assumption that semantic features of the head percolate upward to the phrasal level; in particular, semantic features of the verb are assumed to percolate upward to determine the semantic features of the sentence (this is made explicit in the Head Feature Convention of GPSG and HPSG, and in the [$\uparrow = \downarrow$] feature of heads in LFG). [10]

This view of the principle of compositionality can be shown to be inadequate. More substantive principles of composition—viewed here as constructions—are needed. This can be demonstrated by the existence of cases in which the requirements of the construction are in conflict with the requirements of the main verb. Two cases are discussed below: the Dutch impersonal passive construction and the English *way* construction.

The Dutch Impersonal Passive Construction

Zaenen (1991) provides an argument for a constructional account of the Dutch impersonal passive. There is a constraint on the impersonal passive that the described situation be atelic:

(20) *Er werd opgestegen.
       There was taken off.

(21)   Er werd gelopen.
       There was run.

(22)*?Er werd naar huis gelopen.
       There was run home.

She notes that the acceptability of the sentence can be altered by the addition of particular adverbs:

(23)   Van Schiphol wordt er de hele dag opgestegen.
       From Schiphol there is taking off the whole day.

(24)   Er werd voordurend naar huis gelopen.
       There was constantly run home.

Thus the constraint on the impersonal passive seems to be a constraint on the aspect of the entire expression, rather than one directly on the Aktionsart of the main verb. However, this being the case, the construction cannot be said to be lexically governed: the constraint must be associated with the construction as a whole.

Recognizing the controversial nature of such a proposal, Zaenen explicitly argues against the alternative move—postulating dual senses of each verb, one telic and one atelic. Her argument is based on the fact that another phenomenon in Dutch, auxiliary selection, crucially relies on the inherent Aktionsart of the main verb and can*not* be altered by adverbial modification. The auxiliary *zijn* is chosen when the verb's Aktionsart is telic, regardless of whether the sentential expression is telic or atelic:

(25)   Hij is opgestegen.
       It has taken off.

(26)   Hij is dagelijks opgestegen.
       It has taken off daily.

The auxiliary *hebben,* on the other hand, is chosen when the verb's Aktionsart is atelic. A theory which posited two lexical items, with opposite Aktionsart specifications, would not be able to predict these facts about auxiliary selection. One could conceivably add further features to the description of the main verbs, but such a move would only be motivated by the desire to avoid recognizing the effect of contextual factors independent of the verb. A more satisfactory solution is to posit a single verb sense and allow the impersonal passive to be sensitive to factors outside the main verb.

The *Way* Construction

Another example arises from the constraints on the *way* construction, exemplified in (27) and discussed in chapter 9.

(27) a. Pat fought her way into the room.
     b. Volcanic material blasted its way to the surface.
     c. The hikers clawed their way to the top.

Levin & Rappaport Hovav (1992), following Marantz (1992), have argued that the *way* construction is associated only with unergative verbs. At the same time, they have argued that verbs of directed motion are unaccusative (Levin & Rappaport Hovav 1992). On a lexical account, in which syntactic frames are projected from the verbs' lexical semantics, there is an inconsistency here. All verbs appearing in this construction would have to be considered directed motion verbs, since *way* expressions specifically assert motion along the designated path. This would lead one to the conclusion that such verbs are both unergative (since they occur in the *way* construction) and unaccusative (since they are directed motion verbs).[11]

Alternatively, one might postulate a constraint that the verbs involved must be unergative before they undergo a lexical rule which turns them into unaccusative verbs as expressed in this construction. But this would be an odd kind of constraint: one must worry about not only whether the verb is of the relevant kind as the output of the rule, but also whether the verb was derived in a particular way, in order to determine whether it will occur in this syntactic pattern. Typically, if a verb matches the output of a particular lexical rule, then it behaves like other verbs that have undergone the rule, whether or not it underwent the rule itself (see, e.g., Pinker 1989:65ff.). By contrast, given the more complicated constraint needed here, one would need to know the derivational history of a particular item before one could determine whether it could take part in the argument structure of the *way* construction.

By recognizing the existence of contentful constructions, we can save compositionality in a weakened form: the meaning of an expression is the result of integrating the meanings of the lexical items into the meanings of constructions.[12] In this way, we do not need to claim that the syntax and semantics of the clause is projected exclusively from the specifications of the main verb.

1.4.5   Supportive Evidence from Sentence Processing

Certain psycholinguistic findings reported by Carlson and Tanenhaus (1988) suggest that uses of the same "core meaning" of a verb in different syntactic frames do not show the same processing effects that cases of real lexical ambiguity do. For example, notice that *set* truly has two different senses:

(28) a. Bill set the alarm clock onto the shelf.
     b. Bill set the alarm clock for six.

Load, on the other hand, although it can readily appear in the alternate constructions in (29), according to Carlson and Tanenhaus's hypothesis (as well as the current account) retains the same core lexical meaning in both uses:

(29) a. Bill loaded the truck onto the ship.
     b. Bill loaded the truck with bricks.

Carlson and Tanenhaus reasoned that if a reader or hearer initially selects an inappropriate sense of an ambiguous word like *set,* a garden path will result, effecting an increased processing load. On the other hand, if an inappropriate constructional use ("thematic assignment" on Carlson & Tanenhaus's account) is selected, the reanalysis will be relatively cost free since the sense of the verb remains constant and the verb's participant roles ("thematic roles" on Carlson and Tanenhaus's account) are already activated.

Sentences such as those in (28) and (29) were displayed on a CRT, and subjects were asked to decide as quickly as possible whether a given sentence "made sense." It was expected that subjects would anticipate an inappropriate sense of *set* or an inappropriate use of *load* approximately half the time. A theory which posits two distinct senses of *load* to account for the two uses in (29), analogous to the situation with *set* in (28), would presumably expect the two cases to work the same way. Carlson and Tanenhaus found, however, that misinterpreted lexical ambiguity creates a more marked processing load increase than misinterpreted uses of the same verb. The load increase was witnessed by subjects' longer reaction time to decide whether sentences such as (28) involving a true lexical ambiguity made sense, vis-à-vis sentences such as (29), as well as by a marked increase in the number of "no" responses to the question whether a given sentence made sense when a truly ambiguous verb was involved.[13] The data from 28 subjects are presented in the table below (adapted from Carlson & Tanenhaus 1988): mean reaction times in msec to those sentences judged to make sense are given; the percentages of sentences judged to make sense appear in parentheses:

| | Type of verb | |
Type of ambiguity	Ambiguous	Control
Sense (e.g. *set*)	2445 (77%)	2290 (94%)
Variable constructions ("Thematic ambiguity," e.g. *load*)	2239 (92%)	2168 (93%)

When sentences are divided into preferred and non-preferred sense or construction for a given pair of sentences, the difference in reaction times between different senses and different constructions is even more striking:

| | Type of verb | |
	Ambiguous	Control
Sense ambiguity		
Preferred sense	2277	2317
Less-preferred sense	2613	2264
Variable constructions ("Thematic ambiguity")		
Preferred assignment	2198	2177
Less-preferred assignment	2268	2158

This finding is difficult to account for if one holds the view that different uses of a verb actually reflect lexical ambiguities. That is, on such a view it is difficult to distinguish different uses from different senses, since each different use would entail a different sense (and conversely, each different sense would entail a different use). On the other hand, the distinction found between verbs like *set* and those like *load* is not unexpected on the constructional approach proposed here, since it is claimed that different uses of the same verb in various constructions do not entail different senses of the verb. Thus we would not expect the same verb in different constructions to have the same effect as cases of real lexical ambiguity.

1.4.6   Supportive Evidence from Child Language Acquisition

By recognizing that the meanings of verbs do not necessarily change when these verbs are used in different syntactic patterns—that the meaning of an expression also depends on the inherent semantics of the argument structure constructions—certain findings in language acquisition research can be made sense of.

Landau and Gleitman (1985) note that children acquire verb meanings with surprising ease, despite the fact that the situations in which verbs are used only constrain possible meanings to a very limited degree (cf. also Quine 1960). For example, they note that their congenitally blind subject learned the meanings of *look* and *see* without undue difficulty, despite the fact that these meanings are nonphysical and, for this child, not directly experientially based. They propose that children rely on syntactic cuing, or *syntactic bootstrapping,* as they acquire verbal meaning. In particular, they argue that children make use of the set of syntactic frames that a verb is heard used with in order to infer the meaning of the verb. They argue that this is possible because syntactic frames are surface reflexes of verbal meanings: "The allowable subcategorization frames, taken together, often tell a semantically quite transparent story, for they mark some of the logical properties of the verb in question" (p. 140). Further, they assert that the use of a verb in a particular syntactic frame indicates that the verb has a particular component of meaning, one associated with that syntactic frame. Certain experimental work by other researchers substantiates the idea that syntactic frames aid in the acquisition of word meaning (see Brown 1957; Katz, Baker & McNamara 1974; Naigles 1990; Fisher et al. 1991; Gleitman 1992; Naigles et al. 1993).[14]

However, Pinker (1989) rightly criticizes Landau and Gleitman's formulation of the claim. He notes that if different syntactic frames are assumed to reflect different components of the meaning of verbs, as Landau and Gleitman assume, then taking the union of these different components of meaning across different syntactic frames will result in incorrect learning. For example, if the appearance of an *into*-phrase in *The ball floated into the cave* is taken to imply that *float* has a motion component to its meaning, then the child will incorrectly infer that it will not be possible to float without moving anywhere.

This is indeed a general problem for Landau and Gleitman's formulation. The occurrence of *kick* in the ditransitive construction (e.g., *Joe kicked Mary a ball*) cannot be taken as evidence that *kick*'s meaning has a transfer component, as their account would seem to imply. As we saw above in section 1.4.2, *kick* can occur in eight different syntactic patterns, most of which do *not* involve transfer.

Pinker's criticism rules out the possibility that even adult speakers could use the *set* of syntactic frames a verb is heard used with to determine the verb's meaning. It does so because each distinct syntactic frame is taken to reflect a different sense of the verb. This apparent paradox can be resolved by recognizing that syntactic frames are directly associated with semantics, independently of the verbs which may occur in them. Thus it is possible to recognize that to a large extent, verb meaning remains constant across constructions; differences in the meaning of full expressions are in large part attributable directly to the different constructions involved. On this view, *kick* has the same sense in each of the eight argument structures listed in section 1.4.2. The interpretations—

such as, 'X ACTS', 'X ACTS ON Y', 'X DIRECTS ACTION AT Y', 'X CAUSES Y to UNDERGO a CHANGE OF STATE'—are associated directly with the particular constructions involved. In this way, Landau and Gleitman's insight can be slightly reinterpreted. What the child hypothesizes, upon hearing a verb in a particular previously acquired construction, is not that the verb itself has the component of meaning associated with the construction, but rather that the verb falls into one of the verb clusters conventionally associated with the construction (cf. chapter 5).

Hearing a verb used in different constructions may then indeed aid in the acquisition of verb meaning. One way this might be accomplished is by triangulating the verb class that the verb must belong to. For example, if a child hears an unfamiliar verb occur in a particular construction that is known to be associated with, say, eight verb clusters, and the child also hears the verb used in a different construction that is known to be associated with, say, ten verb clusters, only some of which are shared with the former, the child can narrow down the possible class of verbs by examining only the intersecting clusters.

Contextual information is undoubtedly added into the equation, allowing the child to further narrow down the possible verb classes. That is, language learning does not take place in a vacuum. It is generally accepted that children's first understanding of lexical meaning is tied to the situations in which a word is heard used.[15]

Once constructions are recognized, the idea that the syntactic frames a verb is heard in can aid in determining verb meaning is made coherent. However, as it stands, this account presupposes that the child already knows certain verb classes to be conventionally associated with certain constructions; that is, this account presupposes that a fair number of verbs have already been learned, and so would not provide an account of bootstrapping from ground zero. Constructions would be allowed to aid in the acquisition of the meanings of novel verbs once a fair number of verbs had already been learned, but they would not be useful in acquiring the meanings of the first verbs as Landau and Gleitman have proposed.

Constructions could be claimed to play a more central role in the acquisition of verbal semantics if it were possible to delimit a priori the potential range of verb classes that might be associated with a construction. And in fact it seems there are only a handful of ways that verb meaning and constructional meaning can be related (cf. section 2.5). The necessity of triangulating the relevant verb cluster could be avoided then, since the meaning of the verb would be assumed to be related to the meaning of the construction in one of a small number of possible ways. What is crucial is that the verb's meaning need not directly reflect the meaning associated with the construction. The child's task would be to determine whether the verb's meaning in fact did elaborate the meaning of the construction, or whether the verb coded, say, the means, manner, or result associated with the meaning of the construction.[16]

To summarize, by recognizing skeletal syntactic constructions as meaningful in their own right, it is possible to allow for multiple syntactic frames to be used as an aid in the acquisition of verb meaning. This is because it is not necessary to assume that every use of a particular lexical item in a different syntactic frame entails a different sense of the verb involved.

In the following section, traditional motivations for positing lexical rules to account for variability in syntactic expression are discussed, and it is argued that they are ultimately not persuasive reasons for rejecting a constructional approach.

1.5   TRADITIONAL MOTIVATIONS FOR LEXICAL RULES

There are a number of different types of lexical rule accounts which deal with the issue of variability of overt expression. Lexicalists argue that much of the work that had been done by syntactic transformations is better done in the lexicon. For example, they claim that transformations such as passive, causativization, and dative shift are better captured by lexical rules (Freidin 1974; Bresnan 1978; Mchombo 1978; Foley & Van Valin 1984; Marantz 1984; Pollard & Sag 1987, 1994).[17]

One proposed motivation for adopting a lexical approach to alternations is that many alternations seem to be sensitive to lexical items, particularly verbs. The notion of lexically governed rules goes back to Lakoff (1965), who recognized that no alternation seems to be exceptionless, and that the verb involved largely determines whether a given alternation applies or not. He states: "In some sense the verb 'governs' the passive transformation: it is central to the operation of the rule . . . . There are a number of other clear cases where it is obvious which item it is that governs the rules. Most of these involve verbs" (p. 28). However, in a passage immediately following this suggestion of a notion of government, Lakoff candidly recognizes: "Government . . . is not yet a completely well-defined notion, and we can offer no proposal for an adequate definition of it." In point of fact, the verb alone often cannot be used to determine whether a given construction is acceptable. Consider the following examples:

(30)   a.   Sam carefully broke the eggs into the bowl.
       b.   *Sam unintentionally broke the eggs onto the floor. (cf. section 7.5.1)
(31)   a.   This room was slept in by George Washington.
       b.?* This room was slept in by Mary. (Rice 1987b)

(32) a. Joe cleared Sam a place on the floor.
      b. *Joe cleared Sam the floor. (Langacker 1991)

Holding the verb constant, the (a)-sentences are better than the corresponding (b)-sentences. There is no natural way to capture these types of constraints in the lexical semantics of the main verb. On a constructional account, however, it is possible to associate constraints on the complements or on the overall interpretation of the expression directly with the construction. For example, Rice (1987b) argues that prepositional passives such as those in (31) are more felicitous when the surface subject argument is construed as affected. Similarly, the problem with example (32b) can be seen to be that the ditransitive construction implies that the argument designated by the first object comes to receive the argument designated by the second object. In this case Joe doesn't "receive" the floor, whereas in (32a) he does "receive" a place on the floor.[18]

A second motivation often cited for a lexical account stems from the fact that the lexicon is viewed as the receptacle of all idiosyncratic information. Therefore the existence of idiosyncratic properties is taken as evidence for a lexical phenomenon (Jackendoff 1975; Wasow 1977; Dowty 1979). However, if the lexicon is defined as the warehouse of idiosyncratic information, it must contain information about particular grammatical constructions that are phrasal and even clausal. For example, each of the following is idiomatic in the sense that some aspect of its form and/or meaning is not strictly predictable given knowledge of the rest of grammar.

(33) a. Why paint your house purple? (Gordon & Lakoff 1971)
      b. The more you stare at it, the less you understand. (Cf. Fillmore, Kay & O'Connor 1988)
      c. He cried himself to sleep. (Cf. chapter 8)

Therefore evidence that a phenomenon is idiosyncratic is not evidence that it is *lexical,* unless "lexical" is defined so as to describe all and only idiosyncratic items. But once the definition of "lexical" is extended to this degree, the inevitable consequence is that the lexical is no longer neatly delimited from the syntactic (cf. DiSciullo & Williams 1987).

A third motivation is that crosslinguistically, many alternations are accompanied by morphological marking on the verb. For example, applicatives, causatives, and passives crosslinguistically tend to involve overt morphology on the verb stem. The morphological markers are taken to be evidence for a lexical rule that changes the inherent subcategorization (or semantic representation) of the verb stem. However, the approach suggested here can account for these cases without appealing to any type of lexical rule. On the present account, the closed-class grammatical morpheme is analogous to the English skeletal construction; the verb stem plays the role of the main verb. The semantic integration of morpheme and verb stem is analogous to the integration of construction and verb in English. Since morphemes *are* constructions, and since no strict division is drawn between the lexicon and the rest of grammar, the analogy is quite strong. In fact, Emanatian (1990) has proposed an account along these lines for the Chagga applicative morpheme, as has Maldonado Soto (1992) for the Spanish reflexive morpheme *se*.

A final motivation is that "output" verbs undergo word formation processes, which are generally supposed (since Chomsky 1970, Aronoff 1976) to be a result of lexical rules. Because lexical rules and syntactic rules are taken to be independent, and because lexical rules are assumed to be ordered before syntactic rules, evidence that a rule R feeds a lexical rule is taken as evidence that R is a lexical rule. For example, Bresnan (1982) argues that passive must be a lexical rule since the output of passive is the input to a lexical "conversion" rule of adjective formation. The conversion rule takes passive participles and changes them into adjectives, which are then available as adjectival passives; this accounts for the identity of form between verbal and adjectival passives. Given the lexical nature of the conversion rule, Bresnan concludes: "Since it is assumed that the rule systems of natural language are decomposed into components of lexical rules [and] syntactic rules, . . . which are subject to autonomous sets of constraints, this constitutes the strongest possible kind of evidence that Passivization is a lexical rule" (p. 16). However, there is reason to think that the partition between lexical rules and syntactic rules is not so clearcut (cf. Stowell 1981; Sproat 1985; le Roux 1988; Ward, Sproat & McKoon 1991). Even if we do assume that it is possible to neatly divide grammar into separate components, the lexical and the syntactic—an assumption that Construction Grammar explicitly rejects—it is further necessary to assume that these modules must interact serially, and that syntactic phrases can never feed word formation rules, in order for the type of argument given above to be persuasive. But there are in fact cases of phrasal forms that appear to serve as input to word formation processes. Lieber (1988), for example, argues that the following examples involve phrasal forms which act as the input to lexical compound formation: *a punch-in-the-stomach effect, a God-is-dead theology, a thinking-about-it wink, a connect-the-dots puzzle, a win-a-Mazda competition,* and *a stick-it-in-your-ear attitude* (pp. 204–205).

Thus traditional motivations for accounting for variable syntactic expression in terms of lexical rules are ultimately not persuasive reasons to reject a constructional approach. In the following chapters, such an approach is outlined in more detail.

2 The Interaction between Verbs and Constructions

The constructional approach to argument structure brings several tricky questions to the fore. If basic sentence types are viewed as argument structure constructions, and we wish to claim that essentially the same verb is involved in more than one argument structure construction, we need to deal with the following questions:

1. What is the nature of verb meaning?
2. What is the nature of constructional meaning?
3. When can a given verb occur in a given construction?

Although I have argued that constructions have meaning independently of verbs, it is clearly not the case that the grammar works entirely top-down, with constructions simply imposing their meaning on unsuspecting verbs. In point of fact, there are reasons to think that the analysis must be both top-down and bottom-up. As will be discussed more fully below, the meanings of constructions and verbs interact in nontrivial ways, and therefore some cross-reference between verbs and argument structures will be necessary.

It might be worthwhile to note that the general idea of invoking two simultaneous mechanisms has been recently challenged by Baker (1987), who argues simply that involving two separate mechanisms as opposed to a single mechanism should make learning more difficult—because some mediation between the two mechanisms would be necessary—and should therefore be dispreferred as a psychologically plausible account.

This view, although having some degree of intuitive appeal, has been shown to be false in other domains of cognitive processing. The clearest evidence comes from the domain of vision. For example, it is well known that the perception of depth does not follow from a single principle but from the integration of information of many kinds. Perhaps the most important mechanism is *stereopsis,* the fusing of the two disparate images from the two retinas into a single image. However, stereopsis alone is not the only mechanism by which we determine depth (as can be demonstrated clearly by closing one eye: the perception of depth remains for the most part intact). Other cues include occlusion and differences in gradients of texture (Gibson 1950).

Another example that demonstrates the need for simultaneous mechanisms, and in particular, both top-down and bottom-up processing, comes from letter recognition tasks. Wheeler (1970) and others have shown that letters are more quickly recognized in the context of a word than in isolation. This indicates that the recognition of the word (top-down processing) aids the recognition of the letters that make it up. At the same time, recognition of individual letters (bottom-up processing) is a prerequisite to recognition of the word. Recent connectionist models have had success in trying to model this type of interactive mechanism (McClelland, Rumelhart & Hinton 1986).

These counterexamples to Baker's argument from the domain of human vision and from word recognition tasks should make it clear that the type of interactive system that is being proposed here has ample precedent and should not be dispreferred on unempirical claims of what is "simpler." However, before we can move on to discuss the meanings associated with constructions and verbs, it is necessary to describe the type of semantics that will be adopted.

2.1   Frame Semantics

Meanings are relativized to scenes.
Charles Fillmore (1977a)

Many researchers have argued that words are not exhaustively decomposable into atomic primitives (e.g., Fodor, Fodor & Garrett 1975; Fodor et al. 1980). However, it is not necessary to conclude that meanings have no internal structure. Instead, it has been argued that meanings are typically defined relative to some particular background *frame* or *scene,* which itself may be highly structured. I use these terms in the sense of Fillmore (1975, 1977b) to designate an idealization of a "coherent individuatable perception, memory, experience, action, or object" (1977b:84).

The point is made in the following passage by Austin:

> Take the sense in which I talk of a cricket bat and a cricket ball and a cricket umpire. The reason that all are called by the same name is perhaps that each has its part—its *own special* part—to play in the activity called cricketing: it is no good to say that cricket *simply* means "used in cricket": for we cannot explain what we mean by "cricket" *except* by explaining the special parts played in cricketing by the bat, ball, etc. (Austin 1940:73)

Consider the difference between *ceiling* and *roof.* The top of a single-story building is a *ceiling* if construed with respect to the interior of a building, but a *roof* if construed with respect to the exterior. Thus a central difference between the two terms is that their background frames are different.[1] Fillmore (1977b) compares *land* and *ground. Land* is used to denote solid ground as opposed to the sea, whereas *ground* also denotes solid ground but as opposed to air. These terms are distinguished, therefore, primarily on the basis of the frames in which they are defined.

Another Fillmorian example is *bachelor,* often defined simply as 'unmarried man'. Fillmore points out that *bachelor* is in fact defined relative to a background frame of cultural knowledge. For that reason, it is odd in many cases to ascribe the term *bachelor* to particular unmarried men. For example, is the Pope a bachelor? Is a gay man a bachelor? Is Tarzan? Is a hermit? Or a recently bar mitzvahed young man? In these cases, whether the term applies or not is unclear, because certain aspects of the background frame in which bachelorhood is defined are not present.

Sometimes the background frame is fairly simple, and yet the same crucial point can be made. Langacker (1987a) gives the example of *hypotenuse,* which can only be defined with reference to a right triangle, which in turn can only be understood by assuming a notion of hypotenuse. Such an example provides a simple case for which to explain the notion of *profiling* (Langacker 1987a, 1991). Differences in profiling correspond to differences in the prominence of substructures within a semantic frame, reflecting changes in our distribution of attention.

While both *hypotenuse* and *right triangle* are defined relative to the same background frame (or "base," according to Langacker's terminology), the meanings of the terms differ in that different aspects of the frame are profiled. The different terms can be characterized by the following Langacker-style representations:

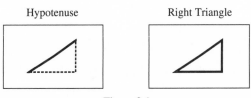

Figure 2.1

Frames in the sense being used here date back to the "schemas" of Bartlett (1932) and have been reintroduced more recently by researchers in Artificial Intelligence including Minsky (1975) and Schank and Abelson (1977). Frames are intended to capture useful chunks of encyclopedic knowledge. Such frame-semantic knowledge has been implemented in FRL (Roberts & Goldstein 1977) and KRL (Bobrow & Winograd 1977) in terms of a hierarchy of data structures (or "frames"), each with a number of labeled slots (see Wilensky 1986 for discussion and critique of various actual implementations; see Gawron 1983 for an application of AI frames to lexical semantics).

Lakoff (1987) argues that certain concepts are defined in terms of a cluster of distinct frames, or "idealized cognitive models." He gives the example of *mother,* which is defined via the following models:

a.   The birth model: the person who gives birth
b.   The genetic model: the female who contributes genetic material
c.   The nurturance model: the female who nurtures and raises a child
d.   The marital model: the wife of the father
e.   The genealogical model: the closest female ancestor

Lakoff argues that the concept *mother* normally involves a complex model in which all of the above basic models are combined. But he notes that oftentimes there is pressure to pick one of the models as criterial, the one that "really" defines the concept, and that which model is picked varies according to circumstance and individual choice:

> I was adopted, and I don't know who my real mother is.
> I am not a nurturant person, so I don't think I could ever be a real
>   mother to any child.
> My real mother died when I was an embryo.
> I had a genetic mother who contributed the egg that was planted in the
>   womb of my real mother (1987:75)

Lakoff goes on to analyze the concept of *mother* as representing a *radial* category: a category with a central subcategory (which in this case combines all of the above models) and noncentral extensions from that prototype (including *adoptive mother, birth mother, foster mother, surrogate mother,* etc.).

2.2   THE NATURE OF VERB MEANING

So far the discussion of frame semantics has centered around nominal examples, but the semantics of nouns and verbs cannot be argued to involve qualitatively different types of knowledge, since nouns are often extended for use as verbs (cf. Clark & Clark 1979). Verbs, as well as nouns, involve frame-semantic meanings; that is, their designation must include reference to a background frame rich with world and cultural knowledge.

It is typically difficult to capture frame-semantic knowledge in concise paraphrase, let alone in formal representation or in a static picture. Still, it is indisputable that speakers do have such knowledge, as a moment of introspection should make clear. Consider the following (oversimplified) definitions:

renege; to change one's mind after previously having made a promise or
    commitment to do something
marry: to engage in a ritualized ceremony with a partner, resulting in a
    change in legal status, with the assumed intention of engaging in con-
    jugal relations and remaining with said partner until one of the two dies

boycott: to avoid buying goods and/or services from a company with the aim of expressing disapproval or causing the company to change one or more of its policies or to go out of business

riot: for three or more people, acting as a group, to engage in activities outside of cultural norms in an unruly and aggressive manner, often with the intention of effecting political consequences

Other examples of verbs requiring aspects of complex world knowledge are not difficult to come by. Consider the rich frame-semantic knowledge necessary to characterize the meanings of: *languish, laminate, saunter, divorce, avenge, promote, subpoena.* In order to capture the richness of these meanings, verbs must be able to refer to conceptual structure, broadly construed (cf. Fillmore 1975, 1977b; Lakoff 1977, 1987; Langacker 1987a; Jackendoff 1983, 1987, 1990a).[2]

The idea that lexical entries should make reference to world and cultural knowledge is not without challengers. While many current theorists using semantic decompositional structures, such as 'X CAUSES Y to RECEIVE Z', 'X ACTS,' or 'X CAUSES Y to MOVE Z', readily recognize that such paraphrases do not capture all of what is intuitively the verb's meaning (e.g., Lakoff 1965; Foley & Van Valin 1984; Levin 1985; Pinker 1989), they argue that such paraphrases are adequate for capturing the "syntactically relevant aspects of verb meaning." The syntactically relevant aspects of verb meaning are defined to be those aspects which are relevant for determining the syntactic expression of arguments via linking rules. Similar proposals have been made by researchers who claim that the theta role arrays associated with lexical entries constitute the only syntactically relevant aspects of verb meaning (e.g., Kiparsky 1987; Bresnan & Kanerva 1989).[3]

On the account proposed here, the semantic decompositional structures correspond to *constructional meanings.* Only in the limiting case do verbs have such skeletal meanings (e.g., *give, do, make*). Since the mapping between semantics and syntax is done via constructions, not via lexical entries, that there *should be* a class of "syntactically relevant aspects of verb meaning" follows from the existence of constructions, which are independently motivated (cf. chapter 1).

Moreover, by distinguishing verbal semantics from constructional semantics, we can predict an observation noted by Pinker as to the nature of "syntactically relevant aspects of verb meaning," or what is here claimed to be constructional meaning. Pinker (1989) observes that such syntactically relevant aspects of verb meaning resemble the meanings of closed-class elements. That is, Pinker notes that the semantic features that are used to predict overt syntactic structure (via linking rules) are the same types of semantic features that have been shown to be associated with closed-class items, for instance motion, causation, contact, and change of state (Talmy 1978, 1983, 1985a; Bybee 1985).

On a constructional account Pinker's observation is predicted. What needs to be recognized is that what Pinker takes to be the "syntactically relevant" aspects of verbal meaning are aspects of constructional meaning. Constructions *are* closed-class elements, so they are predicted to have the semantics of closed-class elements.

A further reason to distinguish the semantics of argument structure constructions from the verbs which instantiate them, and to allow the verbs to be associated with rich frame-semantic meanings, is the need to account for novel uses of verbs in particular constructions. For example, consider the following expression,

(1)   Sam sneezed the napkin off the table.

In order to interpret (or generate) this expression, one needs to know that sneezing involves the forceful expulsion of air. This would not be captured by a skeletal decompositional lexical entry for sneeze such as, for example, 'X ACTS.'

It is also clear that richer aspects of verb meaning are required for aspects of linguistic theory other than predicting the syntactic expression of arguments. For example, frame semantics is needed in order to account for the distribution of adverbs and adjuncts, to account for the process of *preemption* (defined below), to allow for the possibility of meaningful interpretation and translation, and to predict correct inferences. Each of these motivations is discussed in turn.

The particulars of the manner designated by verbs are typically taken to be opaque to syntax (whereas whether the verb encodes a manner or not is taken to be part of the syntactically relevant aspects of verb meaning). For example, with respect to the verb *roll,* Pinker notes: "The idiosyncratic information about the topography of rolling is a black box as far as grammar is concerned, and we need not be concerned about decomposing it, whereas the information that there is a manner specified, or a manner and a path, is something that grammar cares about" (1989:182).

While it may be true that the syntactic expression of arguments is not concerned with specific manners, such specifics are clearly relevant to other aspects of language. In order to account for the distribution of adverbs and adjuncts, reference to the nature of the manner designated by the verb is essential. For example, to predict the distribution of the adverb *slowly,* reference to particulars of manner is required:

(2)   a.   Joe walked into the room slowly.
        b.??Joe careened into the room slowly.

That is, one must know that *careening* implies quick, uncontrolled motion; therefore (2b) is contradictory. Similarly, in order to predict the distinctions between the following examples, reference to the particulars of manner is essential:

(3)   a.   Joe walked into the room with the help of a cane.
       b.   ?Joe marched into the room with the help of a cane.
       c.??Joe rolled into the room with the help of a cane.
       d.   *Joe careened into the room with the help of cane.

Thus the question that is often asked is, what aspects of meaning are relevant for a particular highly circumscribed domain? It is pointed out here that if we wish to ultimately account for a wider domain of language than the syntactic expression of arguments, we will need to appeal to a much richer notion of semantic structure.

Another reason to include frame-semantic knowledge in lexical entries is in order to account for the phenomenon of *preemption*, or "blocking." It is widely recognized that children readily stop using overgeneralized forms upon learning an irregular form with the same meaning. For example, children tend to overregularize *go* to *goed;* but once they realize that *went* is synonymous, they cease to produce *goed.* Thus, *went* is said to preempt *goed.* Similarly, speakers do not generalize the pattern exemplified by *teacher, fighter, listener, doer* to form *cooker*, because *cooker* is preempted by *cook.*

In order for preemption to occur, the hypothesized regular form and the irregular form must have identical semantics. We would not expect *flew* to preempt *soared,* because their meanings are not identical. But in order to determine that *soared* is in fact not synonymous with *flew,* the child must know what *soared* and *flew* mean. It is not enough to know that they are motion verbs with a manner component; the entirety of the frame-semantic knowledge associated with them must be recognized (their phonological dissimilarity is not enough to distinguish them conclusively, since *went* preempts *goed* despite phonetic dissimilarity).

It should also be immediately clear that in order to even have a hope of accounting for interpretation or translation, we need to make reference to frame-semantic knowledge associated with lexical entries. Interpretations that only involve the "syntactically relevant" aspects of verb meaning would leave us with severely underspecified interpretations. For example, consider the following (very) short story:

(4)   Hershel kissed Bolinda. Bolinda slapped Hershel. Hershel slunk away.

This story would be interpreted as:

(5)   Hershel ACTED ON Bolinda in an M_1 manner. Bolinda ACTED ON Hershel in an M_2 manner. Hershel MOVED in an M_3 manner.

We might know that $M_i \neq M_j$, for all $i \neq j$, but clearly such an interpretation is missing an intolerable amount of information. Translation would be rendered impossible, since there would be no means by which to determine correspondence between words.

Finally, it should be obvious that general frame-semantic knowledge is required to account for correct inferences, as has been amply shown by Bartlett (1932), Minsky (1975), Schank and Abelson (1977), and Bobrow and Winograd (1977). To make the case specifically for verbs, contrast the following, for example:

(6)   a.   Sally skipped over the crack in the ground. ($\rightarrow$ she didn't touch the crack)
       b.   Sally crawled over the crack in the ground. ($\rightarrow$ she did touch the crack)

This type of inference is required to determine the acceptability of the following:

(7)   a.   Sally, playing a child's game, avoided touching the crack by skipping over it.
       b.??Sally, playing a child's game, avoided touching the crack by crawling over it.

In order to know whether or not to infer that Sally made contact with the crack, one needs to know exactly what manner of motion is involved in skipping and crawling; the knowledge of the specific manners involved is part of our frame-semantic understanding of what these terms mean. It is not enough to know simply that these verbs encode *some* manner.

To summarize, rich frame-semantic knowledge associated with verbs is necessary for (1) felicitous use of adverbs and adjuncts, (2) interpretation and translation, (3) the process of preemption, or "blocking," and (4) making correct inferences. Unless we decree that the distribution of adverbs and adjuncts, preemption, interpretation, and inferences are not within the domain of grammar, lexical entries must have access to such knowledge.

2.3   THE NATURE OF CONSTRUCTIONAL MEANING
2.3.1 Polysemy

Constructions are typically associated with a family of closely related senses rather than a single, fixed abstract sense. Given the fact that no strict division between syntax and the lexicon is assumed, this polysemy is expected, since morphological polysemy has been shown to be the norm in study after study (Wittgenstein 1953; Austin 1940; Bolinger 1968; Rosch 1973; Rosch et al. 1976; Fillmore 1976, 1982; Lakoff 1977, 1987; Haiman 1978; Brugman 1981, 1988; Lindner 1981; Sweetser 1990; Emanatian 1990). That is, since constructions are treated as the same basic data type as morphemes, that they should have polysemous senses like morphemes is expected. It is worth discussing a particular example of such constructional polysemy.

Ditransitive expressions in English typically imply that the agent argument acts to cause transfer of an object to a recipient. It is argued below that this case of actual successful transfer is the *basic sense* of the construction.

At the same time, it is widely recognized that many ditransitive expressions do not strictly imply that the patient argument is successfully transferred to the potential recipient. For example, a so-called "*for*-dative" expression such as *Chris baked Jan a cake* does not strictly imply that Jan actually received the cake. It may happen that Chris was mugged by cake thieves on the way over to Jan's. In general, expressions involving verbs of creation (e.g., *bake, make, build, cook*) and verbs of obtaining (e.g., *get, grab, win, earn*) do not strictly imply that the agent causes the potential recipient to actually receive the patient argument. Transfer is rather a *ceteris paribus* implication. What is implied by *Chris baked Jan a cake* is that Chris baked a cake *with the intention* of giving the cake to Jan. In fact, many of the verb classes associated with the construction can be seen to give rise to slightly different interpretations.

Expressions involving verbs which imply that the agent undertakes an obligation (e.g., *promise, guarantee, owe*) also do not strictly imply transfer. For example, *Bill promised his son a car* does not imply that Bill actually gave his son a car, or even that Bill intended to give his son a car. Rather, transfer is implied by the "conditions of satisfaction" associated with the act denoted by the predicate (Searle 1983). A *satisfied* promise, for example, does imply that the "promise" receives whatever is promised.

Expressions involving verbs of future having (e.g., *bequeath, leave, refer, forward, allocate, allot, assign*) imply that the agent acts to cause the referent of the first object to receive the referent of the second object at some future point in time. This class differs from the last two classes in that no intention or obligation of future action on the part of the referent of the subject is implied; the agent's role in the transfer is accomplished by the action referred to by the predicate.

Expressions involving verbs of permission (e.g., *permit, allow*) imply merely that the agent *enables* the transfer to occur, by not preventing it—not that the agent actually *causes* the transfer to occur. For example, *Joe allowed Billy a popsicle* implies only that Joe enabled Billy to have a popsicle or did not prevent him from having one—not that Joe necessarily caused Billy to have a popsicle.

Expressions involving verbs of refusal (e.g., *refuse, deny*) express the negation of transfer, for example in *Joe refused Bob a raise in salary* and *His mother denied Billy a birthday cake.* Here transfer is relevant in that the possibility for successful transfer has arisen, but the agent is understood to refuse to act as the causer of it.

Because of these differences, the semantics involved can best be represented as a category of related meanings. That is, the ditransitive form is associated with a set of systematically related senses. Thus the ditransitive can be viewed as a case of *constructional polysemy:* the same form is paired with different but related senses. By accounting for these differences in terms of constructional polysemy, as opposed to positing a collection of lexical rules, for example, we can capture the relations between the different senses in a natural way. In particular, a polysemous analysis allows us to recognize the special status of the central sense of the construction.

The central sense of the ditransitive construction can be argued to be the sense involving successful transfer of an object to a recipient, with the referent of the subject agentively causing this transfer. There are several reasons to adopt this view. The central sense proposed here involves concrete rather than metaphorical or abstract (here: potential) transfer, and concrete meanings have been shown to be more basic both diachronically (Traugott 1988; Sweetser 1990) and synchronically (Lakoff & Johnson 1980). Further, this is the sense most metaphorical extensions (as described in chapter 6) are based on. For example, consider (8) and (9):

(8)   Mary taught Bill French.
(9)   Mary taught French to Bill.

(8) implies that Bill actually learned some French, that the metaphorical transfer was successful. This is in contrast to (9), in which no such implication is necessary. Similarly, compare (10) and (11):

(10)   Mary showed her mother the photograph.
(11)   Mary showed the photograph to her mother (but her nearsighted mother couldn't see it).

(10) implies that her mother actually saw the photograph, whereas for many speakers, no such implication is given in (11).

These facts can be accounted for once we recognize actual successful transfer as the central sense of the construction; we need only state that metaphorical extensions have as their source domain this central sense.[4] Finally, successful transfer is argued to be the central sense because the other classes of meanings can be represented most economically as extensions from this sense.

At the same time, the various senses are not predictable and must be conventionally associated with the construction. For example, it is not predictable from knowing the rest of English that verbs of creation will be allowed in the ditransitive construction in the first place; moreover, it is not predictable that ditransitive expressions involving verbs of creation will imply intended transfer instead of actual transfer or general benefaction. Because of this, the various different possible senses need to be listed.

The suggestion here of allowing for a fairly specific central sense of the construction and postulating separate related senses which make reference to specific verb classes can be contrasted with the possibility of postulating a single abstract sense for the construction and allowing the verbs' semantics to fill out the meaning. Since the latter approach is attractive in being more simple, let me take time to demonstrate why such an abstractionist account fails to adequately account for the data.

Several researchers (e.g., Wierbicka, 1986; Paul Kay, personal communication; Frederike Van der Leek, personal communication) have suggested that there is a simple uniform meaning associated with the ditransitive, namely, that there is some kind of special effect on the first object. It is claimed that the nature of this effect is inferred pragmatically. This proposal is attractive in its elegance, but there are several facts weighing against it. For one, the ditransitive construction does not systematically imply any particular special effect on the first object referent that the corresponding prepositional passive does not imply. Many ditransitive expressions do not entail that the first object referent is affected at all. Moreover, there are pragmatically possible interpretations of "affected" which are not possible interpretations of ditransitive expressions.

To illustrate the first point: there is no noncircular reason to think that the first object is any more affected in the following (a)-cases than in the corresponding (b)-cases:

(12) a. Chris baked Pat a cake.
     b. Chris baked a cake for Pat.
(13) a. Chris promised Pat a car.
     b. Chris promised a car to Pat.
(14) a. Chris kicked Pat the ball.
     b. Chris kicked the ball to Pat.

In fact, there is no obvious definition for "affected" which implies that Pat is necessarily affected in (15):

(15)   Chris baked Pat a cake.

Pat may never receive the cake, and in fact may never even know about it.

In addition, it is not possible to construe the first object as affected in just any pragmatically inferable way. For example, even if we know that there are an agent, a patient, and a goal involved (using the definitions of the thematic roles on, e.g., Kay's account), it is possible to pragmatically infer that the way the goal is affected is by the agent throwing the patient at the goal. However, the following cannot be interpreted this way:

(16)   Pat threw Chris the ball.

(17)   Pat hit Chris the ball.

That is, these examples cannot be interpreted to mean that Pat threw or hit the ball *at* Chris. They can only mean that Pat threw or hit the ball so that Chris would *receive* it—in this case, so that Chris would catch the ball. Consequently, we cannot felicitously say:

(18) #In an attempt to injure Chris, Pat threw Chris the ball.

This fact is unexplained by the abstractionist "affectedness" account.

Another abstractionist analysis that has been offered (Goldsmith 1980) is that the thematic role of the first object be described as *prospective possessor,* thus allowing the semantics to be abstract enough to cover all of the possible interpretations of transfer—actual, intended, future, or refused. However, this suggestion as well, and in fact more generally, any abstractionist account, is subject to several criticisms.

One problem is that an abstractionist account cannot capture the intuition that the notion of transfer in general, and *giving* in particular, is basic to the construction, since by virtue of positing only a single very abstract sense, *all* instances instantiate the construction equally. *Give,* however, is the most prototypical ditransitive verb because its lexical semantics is identical with what is claimed here to be the construction's semantics. This intuition seems to be strong enough to be worth worrying about. In fact, I performed an informal experiment to gauge the strength of the intuition that *give* codes the most basic sense of the construction. I asked ten nonlinguists what the nonsense word *topamased* meant in the following sentence:

(19)   She topamased him something.

A full six out of ten subjects responded that *topamased* meant "give." This fact cannot be attributed simply to effects of general word frequency because there are several other words allowed in this construction that are more frequent than *give*. Thus, according to Carroll, Davies, and Richman's (1971) *Word Frequency Book,* which used a 5,000,000-word corpus, *give* occurred 3,366 times in that corpus, while *tell* occurred 3,715 times, *take* 4,089 times, *get* 5,700 times, and *make* 8,333 times. Of these other words, only *tell* was given as a response in my survey, and it was only given by one speaker. None of the other words were given. One might raise the objection that while *give* is not the most frequently occurring word overall, it is nonetheless the most frequently occurring word *in this construction.* However, the point of the experiment was exactly to test whether speakers were aware of the close relation between *give* and the ditransitive construction; the results seem to indicate that they are.

A related problem stems from the fact that not all ditransitive expressions are equally acceptable. There are certain benefactive ditransitives, to be described in section 6.3.4 in terms of a systematic metaphor, which are acceptable to varying degrees for different speakers. Examples of this type include (Green 1974):

(20)   Hit me a home run.

(21)   Crush me a mountain.

(22)   Rob me a bank.

These expressions are severely restricted in their use, as pointed out by Oehrle (1976). Oehrle observes that they are noticeably more felicitous as commands:

(23)   a.   Hit me a home run.
       b.   ?Alice hit me a home run.

And, they are more acceptable when the recipient is referred to by a pronoun. Contrast (23a) with (24):

(24) ?Hit Sally a home run.

On the present account, we can understand these cases to be a limited extension of the basic sense; we do not need to put them on a par with other ditransitive examples, yet we can still treat them as related to the rest of the ditransitives. However, on an abstractionist account, we have to choose whether to include them as ditransitives or exclude them from the analysis. If we include them, we have no way to account for their marginal status and the special constraints they are subject to. If we exclude them, we fail to capture the obvious similarity they bear to other ditransitives, both in their syntax and in their semantics.

Another problem is that it is not predictable that verbs of creation will combine with the ditransitive to imply intended, instead of actual or future transfer. For example, consider (25):

(25)   Chris baked Mary a cake.

This sentence can only mean that Chris baked the cake with the intention of giving it to Mary. It cannot mean that Chris necessarily gave, or will give, the cake to Mary.

Finally, an abstractionist analysis does not readily allow us to account for the fact, mentioned previously, that metaphorical extensions are based on actual transfer, not potential or intended transfer (but cf. note 4 again). That is, if we only postulate an abstract constraint on the first object position, we have no natural way of accounting for the fact that the metaphorical extensions imply this first object to be an *actual* recipient, not a prospective recipient or goal. However, on our constructional polysemy account we can say that the metaphorical extensions have as their source domain the central sense of actual transfer.

These problems arise for any abstractionist account; therefore, any such account can be seen to be unsatisfactory. Instead, a polysemous semantics is warranted. The related senses of the ditransitive construction can then be diagrammed as in figure 2.2. Each of the links extruding from the central sense in this diagram can be motivated by showing that the same relation holds in other areas of the grammar. In fact, remarkably similar patterns of polysemy are shown to exist for the caused-motion construction discussed in chapters 3 and 7. The related senses involve a category of force-dynamically related types of causation as has been described by Talmy (1976, 1985b) and Jackendoff (1990a).

It might be tempting to think that by positing constructional polysemy, we are simply adding complexity to the construction which would otherwise be attributed to the verb. That is, it might be thought that while we avoid polysemy of lexical items by not postulating separate input and output senses of verbs that undergo lexical rules, we create polysemy of the construction instead.

However, that is emphatically not the case. The polysemy attributed to constructions is polysemy that exists independent of our decision as to how verb meanings should be represented, since it corresponds to polysemy across *outputs* of what is generally taken to be a single lexical rule on traditional accounts. For example, the ditransitive construction is typically captured by a single lexical rule which creates a new verb sense, 'X CAUSES Y to RECEIVE Z'. However, we have seen that ditransitive expressions do not necessarily imply 'X CAUSES Y to RECEIVE Z', but may merely imply 'X INTENDS to CAUSE Y to

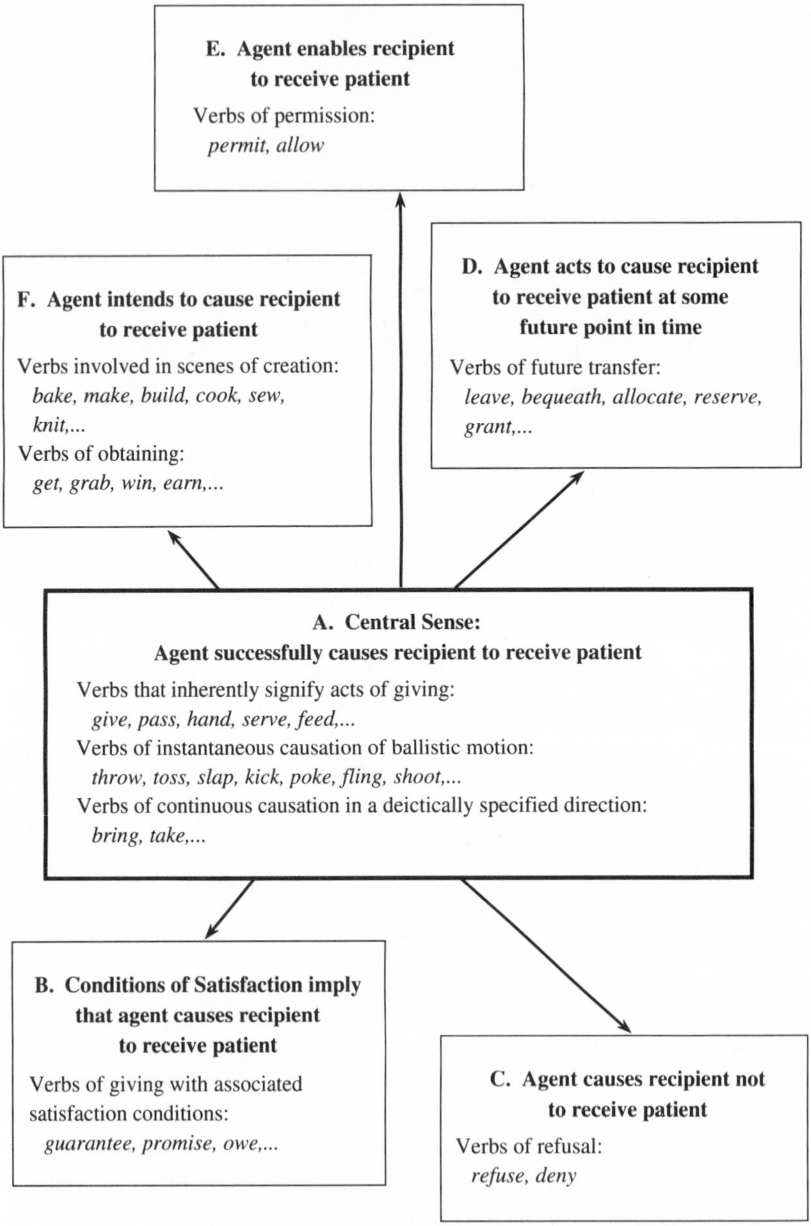

**E. Agent enables recipient
to receive patient**

Verbs of permission:
permit, allow

**F. Agent intends to cause recipient
to receive patient**

Verbs involved in scenes of creation:
*bake, make, build, cook, sew,
knit,...*
Verbs of obtaining:
get, grab, win, earn,...

**D. Agent acts to cause recipient
to receive patient at some
future point in time**

Verbs of future transfer:
*leave, bequeath, allocate, reserve,
grant,...*

**A. Central Sense:
Agent successfully causes recipient to receive patient**

Verbs that inherently signify acts of giving:
give, pass, hand, serve, feed,...
Verbs of instantaneous causation of ballistic motion:
throw, toss, slap, kick, poke, fling, shoot,...
Verbs of continuous causation in a deictically specified direction:
bring, take,...

**B. Conditions of Satisfaction imply
that agent causes recipient
to receive patient**

Verbs of giving with associated
satisfaction conditions:
guarantee, promise, owe,...

**C. Agent causes recipient not
to receive patient**

Verbs of refusal:
refuse, deny

Figure 2.2

RECEIVE Z' (*leave, grant*); alternatively, it may be the case that only the conditions of satisfaction associated with the act designated by the verb imply 'X CAUSES Y to RECEIVE Z' (*promise, owe*) or 'X CAUSES Y not to RECEIVE Z' (*deny, refuse*). Thus on a lexical rule account, a family of lexical rules, each with a slightly different output, would need to be postulated. We may conclude that irrespective of whether we posit distinct verb senses or whether we attribute the resulting semantics to an interaction of verb and construction, it is necessary to account somehow for the observed differences in the resulting semantics.

2.3.2   Humanly Relevant Scenes

In the previous section, it was argued that the English ditransitive construction has as its central sense "successful transfer"—someone causes someone to receive something. In fact, each of the basic clause-level constructions to be discussed can be seen to designate a humanly relevant scene, for example, something causing something to change location (the caused-motion construction), an instigator causing something to change state (the resultative construction), or an instigator moving despite difficulty (the *way* construction). Thus we can form the following hypothesis:

> *Scene Encoding Hypothesis:* Constructions which correspond to basic
> sentence types encode as their central senses event types that are basic
> to human experience.

Languages are expected to draw on a finite set of possible event types, such as that of someone causing something, someone experiencing something, something moving, something being in a state, someone possessing something, something causing a change of state or location, something undergoing a change of state or location, someone experiencing something, and something having an effect on someone. These event types are quite abstract. We do not expect to find distinct basic sentence types which have as their basic senses semantics such as something turning a color, someone becoming upset, someone oversleeping.

The idea that constructions designate scenes essential to human experience is reminiscent of Fillmore's original motivation for the existence of a particular, fixed set of case roles: "The case notions comprise a set of universal, presumably innate, concepts which identify certain types of judgments human beings are capable of making about the events that are going on around them, judgments about such matters as who did it, who it happened to, and what got changed" (1968:24).

Particular combinations of roles which designate humanly relevant scenes are associated with argument structure constructions, which therefore serve to carve up the world into discretely classified event types. Verbs, on the other hand, are associated with richer frame-semantic meanings. As discussed in chapter 5, some cross-reference between verbs and constructions is also necessary, so verbs will in effect be annotated with information about which event types they can be associated with.

In this vein, Langacker (1991) argues that language is structured around certain *conceptual archetypes:* "Certain recurrent and sharply differentiated aspects of our experience emerge as archetypes, which we normally use to structure our conceptions insofar as possible. Since language is a means by which we describe our experience, it is natural that such archetypes should be seized upon as the prototypical values of basic linguistic constructs" (pp. 294–95). He goes on to suggest that these archetypes are extended in various ways for the following reason: "Extensions from the prototype occur . . . because of our proclivity for interpreting the new or less familiar with reference to what is already well established; and from the pressure of adapting a limited inventory of conventional units to the unending, ever-varying parade of situations requiring linguistic expression" (p. 295).

Support for the hypothesis that the central senses of argument structure constructions designate scenes which are semantically privileged in being basic to human experience comes from certain language acquisition facts. In particular, verbs that lexically designate the semantics associated with argument structure constructions are learned early and used most frequently (Clark 1978); certain grammatical markers are applied earliest to "prototypical" scenes—that is, scenes which are claimed to be associated with the central senses of constructions (Slobin 1985); and children's first utterances are about the particular scenes claimed to be associated with constructions (Bowerman 1989). Each of these pieces of evidence is discussed in turn.

Clark (1978) observes that "general purpose verbs" such as *go, put, make, do,* and *get* are often among the first verbs to be used. These verbs designate meanings that are remarkably similar to the meanings associated with argument structure constructions. For example, *go* has the meaning associated with the intransitive motion construction; *put* has semantics very close to that of the caused-motion construction; *make* has the semantics associated with the resultative construction. Possible constructions that are correlated with the meanings of the other high frequency verbs are not explicitly discussed here, but *do* could be said to correspond to the meaning associated with the basic sense of the simple intransitive and/or simple transitive construction. *Get* may well code the semantics of yet another construction, that instantiated by verbs such as *receive, have, take.*

Clark cites other studies which have shown that words corresponding to these concepts are among the first to be used crosslinguistically as well (e.g., Bowerman 1973 for Finnish; Grégoire 1937 for French; Sanches 1978 for Japanese; and Park 1977 for Korean). Children appear to be using these verbs with a general meaning close to that of adults. Clark provides the following interpretations for children's early uses in her data:

do: "perform an action," generally occurring with an agent noun phrase and
    sometimes with an additional patient argument
go: "move," often accompanied by a locative phrase or particle
make: "construct," "produce," or "cause some state to come into being or be
    produced"
put: "cause to be or go in some place"

Not only are these general-purpose verbs learned early crosslinguistically, they are also the most commonly used verbs in children's speech. Clark cites the raw tabulations of verbs used by four children whose mean length of utterance was 2.5 words, collected by Bloom, Miller, and Hood (1975) and Bloom and Lahey (1978) from a fixed database. The data she presents are given in table 2.1. Clark concludes that "*go, put, get, do,* and *make* (plus *sit*) are far more frequent than any other verbs" (1978:48).[5] The fact that these "light verbs," which are drawn from a small set of semantic meanings crosslinguistically, are learned earliest and used most frequently is evidence that this small class of abstract meanings is cognitively privileged. These are the particular meanings directly associated with argument structure constructions.

Table 2.1
Uses of first verbs in fixed data bases (Clark 1978)

Action	Tokens	Locative Action	Tokens
get	*252*	*go*	*417*
do	*169*	*put*	*287*
make	*132*	*sit*	*129*
read	*86*	*fit*	*65*
play	*85*	*take*	*48*
find	*69*	*fall*	*30*
eat	*60*	*go bye-bye*	*28*
fix	*59*	*away*	*26*
draw	*52*	*come*	*25*
hold	*50*	*get*	*25*

It has also been observed that children with Specific Language Impairment rely heavily on the same set of light verbs, including *go, get, do, put,* and *make,* in their production of sentences. For example, sentences such as *I'm doing two balloons* commonly replace the more appropriate *I'm using/playing with/ bouncing two balloons* (Rice & Bode 1993; cf. also Watkins, Rice & Moltz 1993).

Slobin (1985) observes that children's first uses of certain grammatical markings are applied to "prototypical scenes": "In Basic Child Grammar, the first Scenes to receive grammatical marking are "prototypical," in that they regularly occur as part of frequent and salient activities and perceptions, and thereby become organizing points for later elaboration . . ." (p. 1175). He illustrates this claim by arguing that the grammatical marking of transitivity is first used to describe what he terms the "manipulative activity scene." This scene corresponds to the experiential gestalt of a basic causal event, in which an agent carries out a physical and perceptible change of state in a patient by means of direct manipulation. That is, markers of transitivity, both object markers in accusative languages and subject markers in ergative languages, are first applied to the arguments of verbs involving direct physical action, e.g., *give, grab, take, hit,* and not to those of verbs such as *say, see, call out.* In Kaluli (Schieffelin 1985) children do not overextend ergative inflection to the subjects of intransitive verbs, not even when these verbs have an active meaning, like *run, jump;* Slobin thus concludes that children are not grammaticizing the notion of actor in general, but are grammatically marking manipulative activity scenes.

While Slobin considers the acquisition of grammaticalized morphemes, his observations directly carry over to the lexically unfilled constructions in English which are studied here: the morphemes that mark transitivity in other languages correspond to the English skeletal transitive construction, although the latter has no overt morphological marking.

Bowerman (1989) observes more generally that the content of children's first utterances revolves around the general concepts claimed to be associated with constructions: "Regardless of the language being learned, children's first sentences revolve around a restricted set of meanings to do with agency, action, location, possession and the existence, recurrence, nonexistence, and disappearance of objects (Bloom 1970, Bowerman 1973, Brown 1973, Schlesinger 1971, Slobin 1970)" (p. 137). Thus we may conclude that data from language acquisition gives us some independent evidence for the claim that the events encoded by constructions are in some sense basic to human experience.

If it is correct that syntactic ("subcategorization") frames are associated directly with meanings, then what children learn when they learn the syntax of simple sentences is the particular way certain basic scenarios of human experience are paired with forms in their language. That is, we assume that children have already mastered the concepts of transfer between an agent and a willing recipient, causation of motion or change of state, and so forth, and that they come to the task of learning language trying to learn how to encode these basic concepts. Constructions are then extended in various ways allowing the speaker to apply the familiar pattern to new contexts in principled ways, as we saw in the previous section. These patterns of extension are further discussed in the following chapters as well.

At the same time, it is not being claimed that *all* clause-level constructions encode scenes basic to human experience. Nonbasic clause-level constructions such as cleft constructions, question constructions, and topicalization constructions (and possibly passives) are primarily designed to provide an alternative information structure of the clause by allowing various arguments to be topicalized or focused. Thus children must also be sensitive to the *pragmatic information structure* of the clause (Halliday 1967) and must learn additional constructions which can encode the pragmatic information structure in accord with the message to be conveyed. These cases are not discussed further here (cf. Lambrecht 1987, 1994).

2.4   THE INTEGRATION OF VERB AND CONSTRUCTION
2.4.1 Participant Roles of Verbs

Part of a verb's frame semantics includes the delimitation of *participant roles*. Participant roles are to be distinguished from the roles associated with the construction, which will be called *argument roles*. The distinction is intended to capture the fact that verbs are associated with frame-specific roles, whereas constructions are associated with more general roles such as agent, patient, goal, which correspond roughly to Fillmore's early case roles or Gruber's thematic roles.[6] Participant roles are instances of the more general argument roles and capture specific selectional restrictions as well.

A useful heuristic for determining the basic meaning of a verb is to interpret the verb in gerundial form in the following frame:

No _____ing occurred.

The number and type of participant roles implicitly understood to be involved in the interpretation of this expression correspond to the number and type of participant roles in the frame semantics associated with the verb. For example:

(26) a. No kicking occurred.   (two-participant interpretation)
     b. No sneezing occurred.   (one-participant interpretation)

   c. No rumbling occurred.   (one-participant [sound emission] inter-
                                    pretation)
   d. No hammering occurred.   (one-participant [sound emission] or two-
                                    participant [impact] interpretation)
   e. No painting occurred.   (two-participant interpretation—either cre-
                                    ation or coloring interpretation)
   f. No giving occurred   (three-participant interpretation)

In some cases, the verb cannot be used in this frame unless accompanied by
certain complements:

(27) a. *No putting occurred.
          No putting of cakes into the oven occurred.
      b. *No devouring occurred.
          No devouring of cupcakes occurred.

In these cases, the necessarily expressed complements are taken to correspond
to participants associated with the verb.[7]

   Notice that several of the above examples have more than one interpretation,
indicating more than one verb sense. We know from extensive studies of polysemy that lexical items are typically associated with a *set* of related meanings
rather than a single abstract sense (Austin 1940; Wittgenstein 1953; Bolinger
1968; Rosch 1973; Rosch et al. 1976; Fillmore 1976, 1982; Lakoff 1977, 1987;
Haiman 1978; Brugman 1981, 1988; Lindner 1981; Sweetser 1990). Therefore
the existence of two, three, or more distinct but related verb senses is expected.
These polysemous senses can be explicitly related by appealing to the frame
semantics associated with each of them. What is avoided, though, is a system
where a new sense is posited in an unrestrained way for each new syntactic
configuration that is encountered.

Lexical Profiling of Participants

   As was the case with nouns, verbs lexically determine which aspects of
their frame-semantic knowledge are obligatorily profiled. Lexically profiled
roles are entities in the frame semantics associated with the verb that are obligatorily accessed and function as focal points within the scene, achieving a special degree of prominence (Langacker 1987a). These profiled participant roles
correspond to those participants which are obligatorily brought into perspective, achieving a certain degree of "salience" (Fillmore 1977b). Profiling is
lexically determined and highly conventionalized—it cannot be altered by
context.

   In some cases differences in profiling capture the primary difference between verbs. Fisher et al. (1991) appeal to a process that corresponds to profil-

ing to distinguish *take* and *give*. They note: "Movie directors make an art of distinguishing such notions visually. They can zoom in on the receiver's grateful mien, the giver out of focus, or off the frame completely. Using the word *take* rather than *give* is a linguistic way of making the same distinction" (1991 : 8). Similar examples of verbs which seem to invoke the same semantic frame but differ in the participant roles profiled include *loan/borrow, buy/sell* (see Fillmore 1977b for discussion), and *substitute/replace* (see Landau & Gleitman 1985 for discussion).

The test for profiled status that will be used here is that profiled participant roles are those roles which are normally obligatorily expressed in finite clauses. The "normally"-caveat is intended to allow for two types of exceptions: cases where the verb occurs in a construction which serves the purpose of avoiding the overt expression of a particular argument, for example, a passive or middle construction, and (2) cases in which the profiled argument may be unexpressed under certain identifiable contextual circumstances. These two possibilities are discussed in section 2.4.5.

An Example: *Rob* vs. *Steal*

Rob and *steal* at first glance appear to be synonymous, despite their differing syntactic realizations:

(28) a.   Jesse robbed the rich (of all their money).
      b. *Jesse robbed a million dollars (from the rich).[8]
(29) a.   Jesse stole money (from the rich).
      b. *Jesse stole the rich (of money).

However, the differences in the expressions of their arguments can be accounted for by a semantic difference in profiling. In the case of *rob,* the target and the thief are profiled, while in the case of *steal* the valuables and the thief are profiled. Representing profiled participant roles in **boldface,** we might express the difference between *rob* and *steal* thus:

> rob <**thief target** goods>
> steal <**thief** target **goods**>

The different syntactic realizations of participant roles will be shown to follow from differences in profiling, since profiled participant roles must be fused with argument roles that are realized as direct grammatical functions (how this is done is discussed in section 2.4.2).

It might be objected that this putative semantic difference is only postulated to hide an idiosyncratic syntactic difference in the expression of participants. That is, it might be argued that we are only accounting for the fact that the goods role of *steal* and the target role of *rob* must each be linked to OBJ. Indeed, if we needed to stipulate profiling differences as entirely idiosyncratic aspects of lexical entries in order to predict the syntactic expression of arguments, lexical profiling could not be considered a great advance over stipulating the syntactic expression of arguments directly. Either way we would have a stipulation, the only difference being that one stipulation would be semantic and the other syntactic. The differences between the two accounts could then be represented thus:

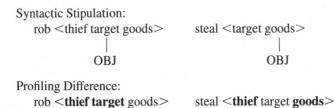

Syntactic Stipulation:
    rob <thief target goods>   steal <target goods>
          | |
       OBJ   OBJ

Profiling Difference:
    rob <**thief target** goods>   steal <**thief** target **goods**>

However, it can be demonstrated that *rob* and *steal* do in fact differ semantically, and that this difference allows us to predict a difference in profiling. *Rob* necessarily entails that the robbed person is seriously negatively affected; this is not true of *steal.* Notice the contrast between (30a) and (30b):

(30) a. I stole a penny from him.
      b. *I robbed him of a penny.

If the victim is indeed negatively affected by the theft, however, use of *rob* becomes acceptable, as can be seen in the following sentence:

(31) I robbed him of his last penny.

Similarly, (32a), in which a rather serious negative effect on the victim is implied, is acceptable, while (32b), in which the effect on the victim is not necessarily serious, is unacceptable:

(32) a. I robbed him of his pride/his livelihood/his nationality.
      b. *I robbed him of his money/a lock of his hair.

Steal, on the other hand, does not require any effect on the victim.

(33) I stole a penny/money/a lock of his hair from him.

Steal focuses on the fact that the stolen goods are not legitimately the thief's property, rather than the fact that they are actually someone else's. The victim is often left vague or unknown:

(34) He stole jewels for a living.

Pinker (1989:396) provides an example which exploits this semantic distinction between *rob* and *steal:*

(35) "She could steal but she could not rob" (The Beatles, "She Came in through the Bathroom Window")

This line plays on the fact that *rob* profiles the victim while *steal* profiles the stolen goods. A person capable of stealing, but not robbing, is deemed relatively less criminal since stealing focuses on the stolen goods and not the victim.

An analogous difference exists between the nominal counterparts of these verbs, *robbery* and *theft* ("the act of stealing"). Robbery is a more serious offense than theft because it entails that the crime is committed *against* someone: the victim has to be present.[9] This is not true of theft. The difference is exemplified in the following:

(36) They charged her with *robbery/theft for shoplifting a jacket.

(37) With an Uzi, the disgruntled citizen committed many robberies/*many thefts.

Figure 23 sums up graphically the differences in semantics between *steal* and *rob.*

Figure 2.3

A Deeper Explanation

I have claimed that the semantic differences between *rob* and *steal* are equivalent to a difference in profiling. A stronger statement would be to say that the primary distinction is in the verbs' semantic frames, and that this distinction underlies or motivates the difference in profiling. Thus it might be argued that the scenes associated with *rob* and *steal* are distinguished by more than a difference in profiling. One piece of evidence for this is the fact that the target role of *steal* is not required to be a person at all—only a source—as we might expect given its syntactic encoding.

(38)   He stole money from the safe.

The same is not true of *rob:*

(39) *He robbed the safe of its contents.

We might distinguish *rob* and *steal* by distinguishing their semantic frames, and thus their participant roles, as follows:

> rob <**robber victim** goods>
> steal <**stealer** source **goods**>

Participant roles such as "victim," which imply direct affectedness by the action denoted by the verb, are instances of the more general role "patient," which is a prime candidate for profiled status across lexical items and across languages. "Source," on the other hand, is rarely lexically profiled, although occasional examples with apparent lexical profiling do exist, as is the case of *depart.*[10]

There are certain generalizations about what types of participants are generally profiled. In particular, participants which are instances of the more general categories "agent" or "patient" tend to be the best candidates for profiled status. Citing Greenfield and Smith (1976), Clark (1978) suggests that agent- or patient-like entities are the most salient to children and are learned earliest: "Most of the object categories named in children's early vocabularies are salient or attractive to them for various reasons: they move on their own, can move other objects, or can be manipulated by children. Notice that they name agents or movers—people and animals. . . . They also name a variety of small-ish objects that are movable or can be manipulated. . . . In contrast, children hardly ever name places, instruments or goals" (1978:35). Fillmore (1977b) also discusses various attributes which tend to cause a participant to be "brought into perspective." Unfortunately a full exploration of the question of which participants tend to be profiled would take us too far afield of the present work, and I do not attempt it here.

2.4.2   Representing the Meaning of Constructions
The Constructional Profiling of Argument Roles

Phrasal constructions, as well as lexical items, specify which roles are profiled. Constructional profiling occurs as follows:

> Every argument role linked to a direct grammatical relation (SUBJ, OBJ, or OBJ$_2$)[11] is constructionally profiled.

The definition of constructional profiling embodies the claim that direct grammatical relations serve to distinguish certain arguments semantically and/or pragmatically; that is, direct grammatical functions profile particular roles as being either semantically salient or as having some kind of discourse prominence, for instance, being particularly topical or focused (see Keenan 1976, 1984; Comrie 1984; Fillmore 1977b; Langacker 1987a, 1991 for arguments to this effect). These grammatical relations are distinguished in most theories as the set of functions which are "terms," or which correspond to "core," "nuclear," or "direct" arguments. Like profiled participant roles, profiled argument roles will be indicated by **boldface.**

It is important to note that the profiling of participant roles discussed above and the profiling of argument roles are not of the exact same kind. The criterion for determining which of a verb's participant roles are profiled is that all and only obligatorily expressed participant roles are profiled. The test for which of a construction's argument roles are profiled is different. In the case of argument roles, all and only roles which are expressed as direct grammatical relations are considered profiled.

Thus the ditransitive construction is associated with the semantics 'X CAUSE Y to RECEIVE Z', which will be represented as

CAUSE-RECEIVE <**agt rec pat**>

The semantics of the construction is expressed in terms of a list of roles simply because this facilitates the statement of the relation between constructional roles and participant roles. However, it should be recognized that neither the constructional roles nor the participant roles constitute an unstructured list of atomic elements. Rather, roles are semantically constrained relational slots in the dynamic scene associated with the construction or the verb (cf. Jackendoff 1983, 1987, 1990a; Foley & Van Valin 1984; Rappaport & Levin 1988; Pinker 1989; Gropen et al. 1991; and Fillmore & Kay 1993 for arguments that roles are not primitives, but are derived from richer semantic structures). Therefore the particular labels that are used to identify these roles have no theoretical significance.

Constructions must specify in which ways verbs will combine with them; they need to be able to constrain the class of verbs that can be integrated with them in various ways (to be discussed in following chapters), and they must also specify the way in which the event type designated by the verb is integrated into the event type designated by the construction. These "principles of integration" between verbs and constructions warrant some discussion.

The Fusion of Participant Roles and Argument Roles

If a verb is a member of a verb class that is conventionally associated with a construction, then the participant roles of the verb may be semantically *fused* with argument roles of the argument structure construction. The term "fusion" is borrowed from Jackendoff (1990a), who uses it to designate the combining of semantic constraints on distinct but coindexed slots within a given lexical entry. I am using the term somewhat differently here, insofar as fusion is meant here to capture the simultaneous semantic constraints on the participant roles associated with the verb and the argument roles of the construction, as opposed to denoting fusion of slots within a single lexical entry. In addition, the possibility of roles fusing is not determined by whether a single role filler can simultaneously fill both roles, but rather by whether the roles themselves are of compatible types.

Which participant roles are fused with which argument roles is determined by two principles:

1. *The Semantic Coherence Principle:* Only roles which are semantically compatible can be fused. Two roles r_1 and r_2 are semantically compatible if either r_1 can be construed as an instance of r_2, or r_1 can be construed as an instance of r_1. For example, the kicker participant of the *kick* frame may be fused with the agent role of the ditransitive construction because the kicker role can be construed as an instance of the agent role. Whether a role can be construed as an instance of another role is determined by general categorization principles.

2. *The Correspondence Principle:* Each participant role that is lexically profiled and expressed must be fused with a profiled argument role of the construction.[12] If a verb has three profiled participant roles, then one of them may be fused with a nonprofiled argument role of a construction.[13] For example, the ditransitive construction can be represented as:

Ditransitive Construction

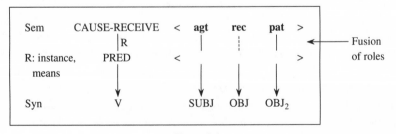

Figure 2.4

The semantics associated directly with the construction is 'CAUSE-RECEIVE <**agt pat rec**>'. PRED is a variable that is filled by the verb when a particular verb is integrated into the construction. The construction specifies which roles of the construction are obligatorily fused with roles of the verb; these are indicated by a solid line between the argument roles and the verb's participant role array. Roles which are not obligatorily fused with roles of the verb—that is, roles which can be contributed by the construction—are indicated by a dashed line. The construction also specifies the way in which the verb is integrated into the construction—what type of relation R can be (see section 2.5 for discussion). Sometimes a specific relation, e.g., means or instances, replaces R in the diagrams below.

Figure 2.4 shows a pairing between a semantic level and a syntactic level of grammatical functions. There is more to say about this linking pattern (cf. chapter 4), but for the moment it is simply stated as a brute force stipulation.

The typical case is one in which the participant roles associated with the verb can be put in a one-to-one correspondence with the argument roles associated with the construction. In this case, the constructional meaning is entirely redundant with the verb's meaning and the verb merely adds information to the event designated by the construction. For example, the verb *hand* is associated with three profiled participants: **hander, handee, handed.** The particular labels of these roles are of no theoretical significance; they are only intended to identify particular participants in the verb's frame semantics.

The three profiled participants of *hand* can be put in a one-to-one correspondence with the profiled argument roles of the ditransitive construction:

Composite Fused Structure: Diransitive + *hand*

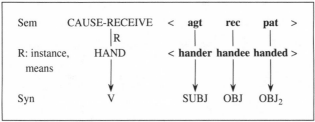

Figure 2.5

The composite structure corresponds to what is traditionally taken to be an additional or derived lexical meaning of the main verb. On the present account, however, the composite structure is just that, a composite structure. Allowing for the constraints specified by individual constructions (which are discussed in some detail in chapters 6–9), new composite structures can be freely constructed.

2.4.3   Mismatches of Roles

Profiling Mismatches

The caused-motion construction is instantiated by expressions such as (40):

(40) Joe squeezed the rubber ball inside the jar.

It can be represented as follows:

Caused-Motion Construction

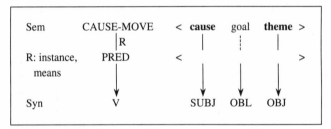

Figure 2.6

Explicit arguments that a construction is required for this case are given in chapter 7.

The participant roles of *put* are fused with the argument roles of the caused-motion construction as follows:

Composite Fused Structure: Caused-Motion + *put*

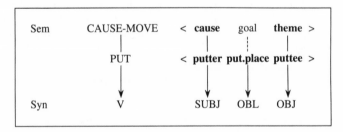

Figure 2.7

In this case, the caused-motion construction's cause argument fuses with the "putter" role of *put,* since a putter is a type of cause. The theme argument fuses with the "puttee," or put-thing, role of *put,* since the roles of theme and put-thing are compatible. The goal (or perhaps more generally, location) argument fuses with the "put.place" role because the "put.place" role is a type of goal.

The goal argument role of the caused-motion construction is not profiled (we can tell because it is linked to an oblique function), although the "put.to" role is (we can tell because it's obligatory). The Correspondence Principle allows for one participant role to be linked to a nonprofiled argument role in cases in which the verb lexically profiles three participant roles. This allows the profiled participant role "put.to" to be fused with the nonprofiled argument role "goal."

The integration of *mail* and the ditransitive construction is an opposite case. *Mail* has three participant roles, two of which are lexically profiled:

    send <**mailer mailed** mailee>

Thus *mail* differs from *hand* in that only two of its participant roles are obligatory:

(41) a. Paul mailed a letter.
      b. *Paul handed a letter.

When *mail* integrated with the ditransitive construction, the construction imposes a profiled status on the "sendee" role:

Composite Fused Structure: Ditransitive + *mail*:

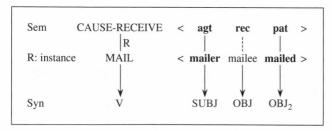

Figure 2.8

In general, if a verb's participant role is fused with a profiled argument role, the participant role inherits the profiled status.

Mismatches in the Number of Roles

Notice that the Correspondence Principle is stated only in one direction: The profiled participant roles must be fused with profiled argument roles (except in the case of three profiled participant roles); that is, all profiled participant roles must be accounted for by the construction. However, it is not

necessary that each argument role of the construction correspond to a participant of the verb. As is argued in more detail in chapters 7–9, the construction
can add roles not contributed by the verb.

For example, the participants of *kick* are kicker and kicked, and the arguments of the ditransitive construction are agent, patient, recipient. The ditransitive construction therefore contributes a recipient role not associated with a
participant role of the verb. The roles are fused as follows:

Composite Structure: Ditransitive + *kick*:

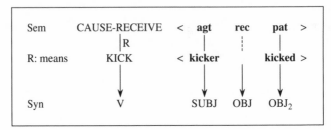

Figure 2.9

The participant roles cannot fuse with the argument roles in any other way
because of the Semantic Coherence Principle. The kicker role can only fuse
with the agent role, because the agent role is the only role it is semantically
compatible with. A kicker is neither a type of recipient nor a patient. The
kicked role is an instance of the patient role but not an instance of the recipient
role.[14] Crucially, the recipient role is contributed by the construction. This
structure yields sentences like (42):

(42) Joe kicked Bill the ball.

Other cases we have seen work similarly. *Sneeze,* for example, has a single
profiled participant role, a sneezer. It integrates with the caused-motion construction as follows:

Composite Structure: Caused-Motion + *sneeze*:

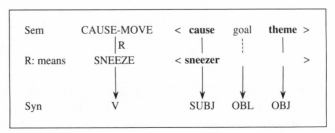

Figure 2.10

The composite fused structure licenses expressions such as (43):

(43) He sneezed the napkin off the table.

Other cases in which constructions contribute roles which do not correspond to participant roles associated directly to the verbs include constructions and verbs exemplified by the following:

(44) "My father *frowned* away the compliment and the insult." (Stephen McCauley, *Easy Way Out,* 1993)

(45) "Sharon was exactly the sort of person who'd *intimidate* him into a panic." (Stephen McCauley, *Easy Way Out,* 1993)

(46) "I cannot inhabit his mind nor even *imagine* my way through the dark labyrinth of its distortion." (Oxford University Press corpus)

(47) Pauline *smiled* her thanks. (Levin & Rapoport 1988)

(48) The truck *rumbled* down the street. (Levin & Rappaport 1990b)

Other Kinds of Mismatches

In all the cases considered so far, the participant roles have been independently classifiable as instances of more general argument roles. However, in other cases, this is not so. For example, consider the verb *send* when integrated into the ditransitive construction. It is assumed that the same sense of *send* is involved in both (49) and (50):

(49) Joe sent Chicago a letter.

(50) Joe sent a letter to Chicago.

The difference in semantics, namely that in (49) Chicago is necessarily construed as standing metonymically for certain people in Chicago, is attributed to an effect of the ditransitive construction, since the construction imposes the constraint that the "send.goal" role must be a recipient, and therefore animate.

The integration of *send* into the ditransitive construction is represented below:

Composite Fused Structure: Ditransitive + *send*:

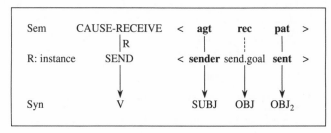

Figure 2.11

Recall that the Semantic Coherence Principle was stated as follows: two roles are semantically compatible iff one role can be construed as an instance of the other. The send goal role can be *construed* as a type of recipient even though it is not necessarily a recipient in and of itself.

Distinct Verb Senses

Occasionally verbs have distinct senses which are systematically related by a difference as to which participant roles are profiled. For example, *lease* and *rent* can occur with either the tenant or the landlord profiled, in addition to the property being profiled:

(51) a. Cecile leased the apartment from Ernest. (**tenant, property**)
     b. Ernest leased the apartment to Cecile. (**landlord, property**)
(52) a. Cecile rented the apartment from Ernest. (**tenant, property**)
     b. Ernest rented the apartment to Cecile. (**landlord, property**)

It might be tempting to think that we could analyze these cases along the lines of the other cases discussed above: we could try to underspecify the meaning of the verb and allow the particular constructions to impose a profiled status on particular roles. In particular, we might try to postulate a single sense of *lease* with the property role as the only lexically profiled participant role. However, our test for profiled participant roles is that all and only roles which are obligatorily expressed in finite sentences are profiled. Given this test, it is not possible to simply say that *lease* only has one profiled role, the property, because the verb cannot occur with only the property role:

(53) *The property leased.

Therefore, to account for these cases, we posit two distinct senses of the verb:

lease$_1$ <**tenant property** landlord>
lease$_2$ <tenant **property landlord**>

Although I have generally tried to avoid positing additional verb senses to account for each possible syntactic pattern, I do not rule out the possibility that *some* alternations must be accounted for by postulating distinct but related verb senses. It should be borne in mind that what we have here is an instance of polysemy, not homonymy, because of the fact that the two senses share the same background semantic frame. They only differ in which roles are profiled.

2.4.4   Unexpressed Profiled Participant Roles

The specific conditions under which a profiled participant role may fail to be expressed are: (1) the verb occurs in a construction which specifically *shades, cuts,* or *merges* the role, or (2) the verb lexically specifies that the role may be unexpressed with a definite interpretation. These topics in themselves could be the subject of a monograph; I do not claim to do them justice here, but I will discuss them briefly in this section.

Shading. The term "shading" is intended to evoke the metaphor suggested by Fisher et al. (1991), that profiling is in some ways analogous to a movie camera focusing on certain participants. Shading denotes a process whereby a particular participant is "put in the shadows," and thus no longer profiled. The passive construction serves to shade the highest ranked participant role associated with the verb. Shading might as well have been termed "deprofiling," except that it is not necessary that the shaded participant is otherwise lexically profiled. Shading is analogous to the suppression of arguments in GB and LFG, although these theories do not make any claims about the semantic/pragmatic effects of passive. A *shaded* participant may be expressed by an adjunct. The statement of the passive requires reference to a thematic hierarchy, versions of which have been proposed, for example, by Fillmore (1968), Jackendoff (1972), Kiparsky (1987), and Grimshaw (1990). The following hierarchy is assumed:

agent, cause > recipient, experiencer > instrument > patient, theme > location, source, goal

The roles expressed by the hierarchy are argument roles, or role types in the sense of Dowty 1986. That is, they are more general than the verb-specific participant roles. Since participant roles are typically instances of one of these roles, the hierarchy serves to define a *partial ordering* of all roles. For example, the "hitter" role is higher on the hierarchy than the "hittee" role. But the fact that the ordering is partial means that not all roles are ordered with respect to each other. Passive applies only to verbs which are associated with two or more roles, one of which is higher than the others.

Passive

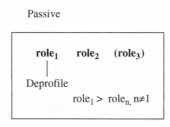

Figure 2.12

Cutting. The term "cutting" is intended to invoke the notion of a director cutting one of the participants out of the picture. Stative constructions in

Bantu (Mchombo 1992), impersonal passive constructions in German, and the middle construction in English serve to *cut* a profiled participant. The difference between a shaded participant role and a cut participant role is that the latter cannot be expressed. For example, the agent role is cut in the English middle construction:

(54) *This bread cuts easily by Sarah.

Role merging. Reflexive constructions, for instance in Romance, serve to *merge* one participant role with another. The merged participant roles are fused with a single argument role, and are then linked with a single grammatical function.[15]

Null complements. Fillmore (1986) distinguishes two distinct ways in which verbs may lexically specify that a certain participant role can fail to be expressed. In the first type of case, the unexpressed role receives an indefinite interpretation; the referent's identity is either unknown or irrelevant. These are *indefinite null complements.* For example, the objects of *eat* and *drink* are not expressed in (53), and their referents' identities—that is, what was eaten or drunk—are irrelevant.[16]

(55)   After the operation to clear her esophagus, Pat ate and drank all evening.

The unexpressed source role in the following is similar:

(56)   Chris drove across the country.

While it is entailed that Chris drove from somewhere, the identity of the source need not be recoverable by either speaker or hearer; it is left indefinite. A similar case involving an unexpressed path argument is given in (57):

(57)   She ran for two hours.

Since the unexpressed role in each of these examples has no special prominence and is nonsalient, these are clear cases of nonprofiled roles. That is, the food and drink participants of *eat* and *drink,* respectively, are participant roles but are not lexically profiled. The same is true of the source (and goal and path) roles of *drive* and *run.*

The second type of unexpressed complement discussed by Fillmore is different: the referent's identity in this case must be recoverable from context. This is the *definite null complement.* Examples of this type include the following (the square brackets are used to indicate where the absent role would normally be expressed):

(58) a. Chris blamed Pat [   ].
      b. Lee found out [   ].
      c. Jo won [   ]!

Only in contexts in which both speaker and hearer can be expected to be able to recover the unexpressed arguments are these cases felicitous; it is in this sense that they are *definite* null complements. Since the contextual constraint ensures that the participant role in question is accessed and salient (in order to be identified), the definite null complement is considered profiled.

Fillmore provides a test to distinguish the two types of unexpressed roles. He notes that while it is perfectly acceptable for a speaker to admit ignorance of the identity of a missing indefinite argument, it sounds odd for a speaker to admit ignorance of a missing definite complement:

(59) a. He found out! #I wonder what he found out. (definite null complement)
      b. He was running. I wonder where he was running to. (indefinite null complement)

Fillmore observes that in English, whether a verb allows an argument to be unexpressed with a definite interpretation is a lexical specification. This assumption is necessary in order to account for distinctions of the following kind:[17]

(60) (Why did you marry her?)
      Because Mother insisted/*required/*demanded. (1986:98)

Only *insist* allows a definite null complement; the closely related *require* and *demand* do not. At the same time, many other languages, including Japanese, Korean, and Hungarian, allow definite null arguments freely. In these languages, often only the verb is overtly expressed: all of the verb's participants may receive a definite interpretation in context. Below, profiled definite omissible participant roles will be represented by the role name in boldface surrounded by square brackets: [**role**].

To summarize, there are several ways in which profiled participant roles can be accounted for without being overtly expressed. The verb may occur in a construction which specifically shades, cuts, or merges a certain role or, in languages like English, the verb may lexically designate that a particular role may be unexpressed if it receives a definite interpretation.

2.5 POSSIBLE RELATIONS BETWEEN VERBS AND CONSTRUCTIONS

On a constructional approach to argument structure, in which the semantics of the verb classes and the semantics of the constructions are integrated to yield the semantics of particular expressions, the question arises as to what range of verb classes can be associated with a given construction.

Could *any* verb class in principle be conventionally associated with a particular construction?[18] For example, if we accept that the ditransitive construction is directly associated with a particular semantics, roughly, 'X CAUSES Y to RECEIVE Z', then why would it not be possible in principle for, say verbs of mood like *sadden, anger, regret* to be used with the ditransitive construction as in (61) to imply the resulting emotional state?

(61) *Joe angered Bob the pink slip.
      ("Joe gave Bob a pink slip, causing Bob to become angry.")

Obviously we want to rule out such a possibility.

In order to circumscribe the possible types of verb classes that can be associated with particular constructions, we need to examine more closely the types of relations that the verb's semantics may bear to the semantics of the construction.

Commonly, the event type designated by the verb is an instance of the more general event type designated by the construction. For example, consider the use of *hand* in (62):

(62)   She handed him the ball.

Hand lexically designates a type of transfer event; at the same time, transfer is the semantics associated with the ditransitive construction. Another example of this kind is *put,* used as in (63):

(63)   She put the phone on the desk.

Put lexically designates a type of caused-motion event, and caused motion is of course the semantics associated with the caused-motion construction.

Other systematic relations between verbs and constructional meanings have been discussed under the heading of "conflation patterns" (Talmy 1985a). In our terms, conflation patterns correspond to mismatches between the semantics of the verb and the semantics designated by the construction. The mismatches can be of several types.

As had been implicit in much of the generative semantics literature (e.g., Lakoff 1965; McCawley 1973) and has more recently been recognized by Talmy (1985), Levin and Rapoport (1988), and Jackendoff (1990a), verbs which do not directly denote the meaning associated with the construction often denote the *means* by which the action is performed. This is the relation that verbs of ballistic motion bear to the meaning of the ditransitive construction. For example, in (64) kicking is the means by which transfer is effected.

(64)   Joe kicked Bob the ball.
        ("Joe caused Bob to receive the ball by kicking it.")

In the case of causative constructions, the verb designates the *result* associated with the construction. The construction supplies an agent argument which does not fuse with any of the participant roles associated with the verb. For example, consider the Chicheŵa causative morpheme *íts* in (65) (from Alsina & Mchombo 1990):

(65)   Nŭngu   i-na-phík-íts-a   maûngu   kwá kádzīdzi.
        9 porcupine 9s-ps-cook-CAUSE-fv 6 pumpkins to   1 owl.
        'The porcupine had the pumpkins cooked by the owl.'

Alsina (1993) analyzes this morpheme as having the following semantic representation:

(66)   CAUSE <agt pat PRED <...>>

The causative morpheme is thus a construction, into which the verb's semantics (represented by PRED) integrates. This morphological construction is quite analogous semantically to the lexically unfilled English constructions that have been discussed so far. The verb stem and the causative morpheme must integrate, just as the English verb must integrate into the various English constructions.

The Causal Relation Hypothesis

Croft (1991) proposes a general constraint on possible conflation patterns. He suggests that "individual lexical items appear to denote only causally linked events" (p. 160) (see also Matsumoto 1991 for discussion of the centrality of causality in this respect). To illustrate his point, Croft cites the following example adapted from Talmy (1985a):

(67)   The boat sailed into the cave.

He argues that the sailing manner and the implication of motion can only be conflated if the activity of sailing *causes* the motion. That is, the following is unacceptable:

(68) *The boat burned into the cave.

Example (68) cannot mean that the boat entered the cave while burning.[19]
    Croft's claim can be restated in terms of the present account in the following way:

> *Causal Relation Hypothesis:* The meaning designated by the verb and the meaning designated by the construction must be integrated via a (temporally contiguous) causal relationship.

Evidence supporting Croft's claim comes from the distribution of verbs of sound emission with constructions that designate motion. Such verbs can be used freely when the sound is a *result* of the motion and occurs simultaneously with the motion:

(69) a. The wooden-legged man clumped into the room.
     b. The train screeched into the station.
     c. The fly buzzed out of the window.
     d. The truck rumbled down the street. (Levin & Rappaport 1990b)
     e. The elevator creaked up three flights.

For instance, the clumping noise of (69a) is a result of the man's moving. For most speakers verbs of sound emission cannot be used for coincidentally co-occurring (or characteristic) sounds, where no causal relationship is involved:

(70) a. *The bird chirped out of the cage.
     b. *The dog barked into the room.
     c. *The rooster crowed out of the barn.
     d. *The man laughed out of the room.

However, Croft's claim is not sufficient to account for all cases. This brings us to the following section.

Violations of the Causal Relation Hypothesis

There are several types of violations of the Causal Relation Hypothesis that are allowed by particular constructions. The construction exemplified by (71) allows verbs which designate events not causally related, at least to a limited extent (cf. chapter 9).

(71) She kicked her way out of the room.

For example, the following examples from the Oxford University Press corpus involve only the *manner* of motion, not the *means* of motion (cf. Levin & Rapoport 1988; Jackendoff 1990a):[20]

(72) a. " 'I *knitted* my way across the Atlantic,' he reveals."
     b. ". . . without a party to go to, he *nods* and *winks* his way through the set crammed with seaside sing-alongs."
     c. ". . . [anyone] watching would have thought he was *scowling* his way along the fiction shelves in pursuit of a book."

Interestingly, the *way* construction tends to be used with pure manner verbs only when the manner is particularly salient and emphasized. This is reflected in the fact that, not uncommonly, manner cases involve two or three conjoined verbs, as in example (72b).

Returning to verbs of sound emission again, it seems that they can marginally be used in the motion construction when the verbs do not designate a sound resulting from the motion. In particular, if the sound is the means of identifying the path of motion, the expressions seem at least marginally acceptable:

(73) a. ?The police car screamed down the street.
      b. ?The train whistled into the station.

The conative construction exemplified by (74) also permits exceptions to the Causal Relations Hypothesis:

(74) a. Ethel struck at Fred.
      b. Ethel shot at Fred.

In this case the verb designates the *intended result* of the act denoted by the construction. The semantics of the construction can be represented roughly as 'X DIRECTS ACTION AT Y'. That is, Ethel does not necessarily strike Fred, but striking him is the intended result of the directed action. The construction can be represented as follows:

Conative Construction

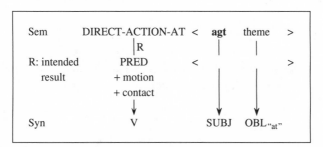

Figure 2.13

The fact that a verb that is related to the construction by the intended-result relation must be [+motion, +contact] serves to allow verbs such as *shoot, hit, kick,* and *cut,* while correctly ruling out verbs such as *move (no contact) and *touch (no motion) (Guerssel et al. 1985; Laughren 1988). This constraint is captured by restricting the class of verbs which can instantiate PRED when the R-relation is one of intended result.

This representation allows us to assimilate expressions such as (74a, b) above to other related expressions, for instance those in (75):

(75) a. Fred looked at Ethel.

b. Ethel aimed at Fred.

Look and *aim* are not [+motion, +contact] verbs,[21] and yet they bear an obvious similarity to the cases above. They differ from these earlier cases in that now the verb's semantics is an instance of the semantics of the construction. That is, 'look' and 'aim' are instances of 'DIRECT-ACTION-AT'. For example, *aim* fuses with the conative construction as follows:

Composite Structure: Conative + *aim*

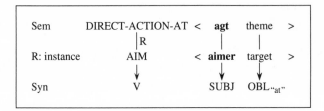

Figure 2.14

The meaning of the construction remains constant, regardless of whether the verb designates an instance or the caused result; it is the relation between the meaning of the verb and the meaning of the construction—the R-relation—which is different. Particular R-relations must be able to refer to classes of verbs in order to capture the [+motion, +contact] constraint. The conative construction can be represented as follows:

Conative Construction

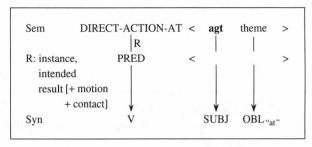

Figure 2.15

Verbs may also code particular *preconditions* associated with the semantics of the construction. For example, creation verbs designate an act of creation, which is a precondition for transfer. Consider (76):

(76) Sally baked Harry a cake.

This sentence does not entail that the baking itself was causally related to the transfer. The baking does not cause the transfer, and the transfer does not cause the baking. However, the creation of the cake is a necessary precondition of the transfer.

An important question is, why should these relations be privileged? Why should means, preconditions, and to a lesser extent, the manner involved in an event be more likely candidates for use in a construction which implies the entire event than, say, the mood of one of the participants?

This deeper question is difficult to answer, but if we consider certain verbs' inherent semantics to bear a *metonymic* relationship to the semantics of the construction, we may find a partial explanation. The semantics associated with the construction defines a semantic frame, and the verb must inherently designate a particular salient aspect of that frame.

The Fusion of Roles

Matsumoto (1991) notes that when two verbs are combined to form a complex motion predicate in Japanese, they must share at least one role. He labels this constraint the Shared Participant Condition. In our terms, this constraint can be translated into the claim that at least one participant role and argument role must be fused; thus not all of the argument roles can be contributed by the construction.

Summary of the Relations between Verb Semantics and Construction Semantics

Let e_c be the event type designated by the construction, and e_v the event type designated by the verb.

I. e_v must be related to e_c in one of the following ways:
   A. e_v may be a subtype of e_c
   B. e_v may designate the means of e_c
   C. e_v may designate the result of e_c
   D. e_v may designate a precondition of e_c
   E. To a very limited extent, e_v may designate the manner of e_c, the means of identifying e_c, or the intended result of e_c

II. e_c and e_v must share at least one participant (Matsumoto 1991).[22]

Do all of the possible relations in (I) have equal status? Clearly not. That e_v may be a subtype of e_c is prototypical and universal. The possibility that e_v may code the means of e_c seems to be a language-specific parameter: English, Dutch, and Chinese allow this relation; Romance, Semitic, and Polynesian languages apparently do not (Talmy 1985a). Other relations, for example that e_v may designate the precondition, manner, or result of e_c, are construction specific.

The result of integrating the verb with the construction must be an event type (E) that is itself construable as a single event. That is, only a single event can be expressed by a single clause. Some of the constraints on exactly what this entails are discussed in chapters 7 and 8.

2.6 CONCLUSION

In this chapter, I have attempted to argue for some of the basic claims underlying this monograph, and have laid out some of the machinery needed to make these claims precise. Following the discussion in chapter 1, where it was argued that constructional meaning exists independently of verb meaning, the type of semantics associated with verbs and constructions has been discussed in more detail.

Verbs and other lexical items have been argued to be associated with rich frame-semantic knowledge. Basic sentence-level constructions, or argument structure constructions, have been argued to designate scenes which are in some sense basic to human experience (cf. also Fillmore 1968, Langacker 1991). That is, it is claimed that the set of basic clause types of a language are used to encode general event types such as those denoting that someone did something to someone, something moved, someone caused something to change state, someone experienced something, someone possessed something, and so forth. Evidence for the idea that these event types have a privileged status comes from certain language acquisition facts noticed by Clark (1978), Slobin (1985), and Bowerman (1989).

In addition it has been argued that these basic senses are extended in various ways so that particular syntactic frames are associated with a family of related meanings. This idea has been explicitly contrasted with the idea that the semantics associated with a construction is ultimately generalized, or that it is abstracted to a single more general sense.

Finally, constraints on the types of potential relations between verbs and constructions have been suggested, extending observations by Talmy (1985a), Croft (1991), and Matsumoto (1991).

3 Relations among Constructions

The repertoire of constructions is not an unstructured set. There are systematic generalizations across constructions. In this chapter, several organizational principles are discussed and applied to the constructions analyzed in this work. It is argued that constructions form a network and are linked by inheritance relations which motivate many of the properties of particular constructions. The inheritance network lets us capture generalizations across constructions while at the same time allowing for subregularities and exceptions.

Before explicating the nature of the relations between the constructions we have looked at, it is important to describe the general psychological principles of language organization that will be assumed.

3.1 RELEVANT PSYCHOLOGICAL PRINCIPLES OF LANGUAGE ORGANIZATION

Each of the following principles is stated in terms of constructions, since constructions are the basic units in our system. All of these principles have direct analogues in various functionalist frameworks.

I. *The Principle of Maximized Motivation:* If construction A is related to construction B syntactically, then the system of construction A is *motivated* to the degree that it is related to construction B semantically (cf. Haiman 1985a; Lakoff 1987). Such motivation is maximized.

II. *The Principle of No Synonymy:* If two constructions are syntactically distinct, they must be semantically or pragmatically distinct (cf. Bolinger 1968; Haiman 1985a; Clark 1987; MacWhinney 1989). Pragmatic aspects of constructions involve particulars of information structure, including topic and focus, and additionally stylistic aspects of the construction such as register (cf. discussion in section 1.1).

> *Corollary A:* If two constructions are syntactically distinct and S(emantically)-synonymous, then they must not be P(ragmatically)-synonymous.
> *Corollary B:* If two constructions are syntactically distinct and P-synonymous, then they must not be S-synonymous.

III. *The Principle of Maximized Expressive Power:* The inventory of constructions is maximized for communicative purposes.

IV. *The Principle of Maximized Economy:* The number of distinct constructions is maximized as much as possible, given Principle III (Haiman 1985a).

In support of these principles, consider the analogy Haiman (1985a) proposes between the form of a language and a diagram such as map or a musical score. Haiman suggests that while a map depicts geography and a musical score depicts a melody, language depicts our construal of reality.

There are several relevant facts about diagrams. Haiman notes that in an ideal diagram, every point should correspond to some point in the reality being depicted. He refers to this property as *isomorphism*, which also seems to imply that every point in the geography or in the melody corresponds to a unique point on the map or musical score, respectively. Moreover, every relation between two points on a diagram should correspond to a relationship between points in reality. This second property is referred to as *motivation*. Haiman notes that working against strict adherence to these two properties is the fact that diagrams are designed to simplify: they only need to represent, not reproduce. For example, a map does not show all of the details of the territory being represented and a musical score diagram does not uniquely determine the way the music is to be played. Moreover, diagrams often contain certain distortions: Greenland is represented too largely in most maps, and a low note in treble clef notation is higher than a high note in bass clef.

However, the general principles of isomorphism and motivation are observed to a large degree: each point of a map corresponds roughly to one point in the world, each representation of a musical note corresponds to only one pitch most of the time. Also, the distance between two points on the map is generally greater when the corresponding distance in the world is greater; within the same clef, higher notes are higher than lower notes.

The analogy to natural language runs as follows. The principle of isomorphism covers two aspects. On the one hand, differences in form imply differences in meaning (or pragmatics), as demanded by the Principle of No Synonymy (cf. section 1.1). Haiman attributes this principle to Humboldt, Vendryes, Ogden, and Richards, and it has been echoed more recently by, among others, Bolinger 1968, Clark 1987, and MacWhinney 1989. Conversely, a difference in meaning or pragmatics should lead to a difference in form, in accordance with what we have called above the Principle of Maximized Expressive Power.

Noting a need for simplification, Haiman allows for derivations from isomorphism. He suggests that deviations from this rule in natural languages occur in cases of polysemy and homonymy, but that such exceptions can be attributed to a general need for simplification, just as in the case of diagrams. This observation is captured by the Principle of Maximized Economy.

Therefore, while the Principle of Maximized Economy works to constrain the multitude of constructions, the Principle of Maximized Expressive Power works in the opposite direction, creating the tendency for more distinct forms; that is, a maximally expressive system would have a distinct label for every distinct item in the user's world. These two principles mutually constrain each other.

With one possible exception, all of the functional principles listed above are widely assumed and are sufficiently intuitive so that a more extended defense of them is not attempted here. The one principle which is somewhat less widely adopted within linguistics is the Principle of Maximized Motivation. Since this principle plays an important role in the discussion of the relations between constructions, it is worthwhile discussing it in more detail.

3.2 MOTIVATION

The term "motivation" was introduced into linguistics by Saussure. In the *Cours* he provides the example of *dix-neuf* 'nineteen', noting that while the parts of this word are arbitrary signs, the complex taken as a whole is *motivated.* It is clear that it is not predictable that *dix-neuf* should take the form it does. A unique morpheme could have been introduced to signify the concept 'nineteen', or *neuf-dix* could have been used. Still, there is an obvious sense in which the term is not arbitrary.

Motivation in this sense lies between predictability and arbitrariness. In an intuitive sense, it often constitutes explanation. If a (somewhat hapless) French child were to ask, "Why is this many [pointing to nineteen things] referred to by '*dix-neuf*'?" a natural response would be to point out that *neuf* means 'nine' and *dix* means 'ten' and that nine plus teen is nineteen.

Haiman argues that making generalizations and simplifications is a necessary function of language, because it would be impossible in our finite world, with our finite memories, to have distinct names for the infinite number of actual distinctions in the world. Rather than recognizing an infinity of sounds and concepts, human language recognizes a finite inventory of phonemes and morphemes. In order to reveal the importance of motivation in grammar, Haiman recounts J. L. Borges's tale of "Funes the Memorious" (1962). Borges's hero, Funes, has undergone an accident that has left him with a perfect memory. Funes can remember "the outlines of the foam raised by an oar in the Rio Negro the night before the Quebracho uprising." Since he has a perfect memory, he devises his own language, in which every sense experience and every concept he recognizes is given a separate name: "It bothered him that a dog at 3:14 (seen from the side) should have the same name as the dog at 3:15 (seen from the front)." Funes scorns the use of mnemonic classification: "In place of

7,013, he would say *Maximo Perez;* in place of 7,014, he would say, *The railroad;* . . . in place of five hundred, he would say *nine.*"

By rejecting principles of organization, Funes's language is not motivated. Every difference is a complete difference; there is no motivation to code generalizations and similarities. It is admittedly often not predictable which generalizations or similarities a language will encode; however, unless the necessity of motivation in a grammar is recognized, we cannot account for the fact that Funes's language is an inconceivable human language.

Langacker (1987a) has also stressed the importance of a notion between predictability and arbitrariness. He notes that our inability to predict what pattern a language uses does not entail that the choice has no semantic basis. For example, he observes that while the fact that *scissors, pants, glasses,* and *binoculars* have the form of plurals is not predictable from their designations, it is nonetheless motivated by the bipartite character of the type of object the words designate (1987a:47).

Lakoff (1987) suggests a precise definition for the term "motivation" in grammar. A given construction is *motivated* to the degree that its structure is inherited from other constructions in the language. On Lakoff's (1987) account of *there*-constructions, the "based-on" relation is of central importance. It is said to be an asymmetric inheritance relation, so that if construction A is based on construction B, then A inherits all of B's properties that do not specifically conflict with its own specifications. Lakoff suggests that the more the properties of a given category are redundant, the more it is motivated and the better it fits into the system as a whole. An optimal system is a system that maximizes motivation. There may be many optimal grammars since motivation can be maximized in many ways.

Researchers in child language acquisition are also arguing against the idea of a strict dichotomy between predictability and arbitrariness. More and more they are advocating learning mechanisms in which there is no sharp division between obligatory rules and probabilistic tendencies (e.g., Bates & MacWhinney 1987; MacWhinney 1989, 1991; Pinker 1987).

Evidence that a relation in form aids in the acquisition of concepts which are related in meaning comes from studies of children's learning of taxonomic relations. Gelman, Wilcox, and Clark (1989) have shown that children learn the names of subordinate terms more easily when those terms are compounded with basic level terms that the child already knows. For example, children were more likely to learn the name for a new type of car when it was called a *fepcar* than when it was simply called a *fep.* This finding is not obvious, since it would seem on the face of it that a child would have to learn more in learning the compound term than in learning the uncompounded novel term. However, when motivation is taken into account as an aide in learning, the findings can be seen to be natural. Children learn new terms for concepts which are related to other, already familiar concepts more easily when the new terms are systematically related to the terms for the familiar concepts.[1]

A recognition of the importance of motivation-like reasoning is growing in the field of Artificial Intelligence. *Abduction,* or reasoning to the best explanation, has been argued to be useful in attempts to model human inferences (Wilensky 1982). Typically one must know the outcome in order to perform abduction, which distinguishes it from *deduction.* In critical respects, the seeking out of linguistic motivation can be understood to be abductive inferencing applied to language learning, whereas predictability corresponds to the result of applying deductive reasoning. That is, abductive reasoning involves after-the-fact inferencing to determine why a given sequence of events should have occurred as it did. The given sequence of events is not, however, a priori predictable. Similarly, while speakers cannot predict whether or to what extent two related concepts will be related formally, it is claimed that they nonetheless search for such relations in order to "make sense of" the input forms, fitting the new forms into the network of interrelated constructions that constitutes their knowledge of language. This idea has been suggested by, for example, Bates and MacWhinney (1987), who propose that relations between forms, meanings, and form–meaning pairs are (unconsciously) observed and pondered in their own right. If Wilensky is right in arguing that people seek out abductive explanations—that is, motivation—in trying to account for sequences of events, then this would give us reason to suspect that speakers might unconsciously apply the same principles in trying to acquire language.

Connectionist representations also make no sharp division between what is predictable and what is arbitrary, instead allowing there to be correlations of varying strength (cf. Rumelhart & McClelland 1986). Individual correlations can be interpreted as motivating factors: they may influence the system in a certain direction but they are not in isolation predictive. Such models have begun to be applied to linguistic phenomena. In these systems violable constraints—constraints which add to the naturalness (or unnaturalness, if framed negatively) of a given expression—are of central importance. This ideas has given rise to constraint optimalization theories in phonology (Smolensky 1986; Legendre, Miyata & Smolensky 1990; Prince & Smolensky 1991; Goldsmith 1993).

More generally, in connectionist networks, items of new information are more easily incorporated when analyzed as variations on known information; new patterns are automatically assimilated to old patterns as much as possible. Optimization in such systems therefore produces motivated structures.

Incorporating motivation into the grammar captures a fundamental structuralist insight which has been overlooked by most formal linguistic theories. This insight is that elements in a system influence each other even when they do not literally interact. Evidence for this kind of influence is abundant in the domain of phonology, for example in the phenomena of analogic extension and restoration, back formations, push and drag chains, paradigmatic leveling, and in the very fact that, to a striking degree, sound change is regular. These phenomena attest to the fact that speakers (unconsciously) seek out regularities and patterns, and tend to impose regularities and patterns when these are not readily available.

The idea of explicitly linking constructions that are related in various ways is in accordance with what is known about the lexicon. Current research overwhelmingly rejects the idea that the lexicon is simply a list of unrelated facts or completely independent pieces of knowledge. Instead, memory in general, and the lexicon in particular, have been shown to involve a richly interconnected web of information. Various psycholinguistic priming experiments have shown that form and meaning relations between lexical items are cognitively real (e.g., Meyer & Schvaneveldt 1971; Ratcliff & McKoon 1978; Anderson 1984).

3.3   REPRESENTING MOTIVATION BY INHERITANCE

To capture relations of motivation, asymmetric *inheritance links* are posited between constructions which are related both semantically and syntactically. That is, construction A motivates construction B iff B *inherits* from A. Inheritance allows us to capture the fact that two constructions may be in some ways the same and in other ways distinct.

The idea of using inheritance as a method of capturing generalizations originated in computer science, as a way to represent data structures in as general a form as possible (cf. Fahlman 1979; Touretzky 1986). Inheritance has since been found to be useful in many programming and knowledge representation systems, including FRL, KRL, KL-ONE, KODIAK, SMALLTALK, FLAVORS, LOOPS, ADA, and object-oriented LISP. By postulating abstraction hierarchies in which lower levels inherit information from higher levels, information is stored efficiently and made easily modifiable.

Use of the concept of inheritance is also currently growing as a way to capture linguistic generalizations, for example in work by Bobrow and Webber (1980), Hudson (1984), Lakoff (1984), Flickinger, Pollard, and Wasow (1985), Wilensky (1986), Pollard and Sag (1987), Jurafsky (1992), Thomason (1992), Davis (1993). The following inheritance system draws on aspects of each of these theories.

Following Lakoff 1984, Wilensky 1986, and Jurafsky 1992, the data structures in our system are constructions. Constructions are specified as to which other, more abstract constructions they inherit from, or equivalently—to use the terminology of Wilensky 1986—which other constructions they are *dominated* by.

3.3.1   General Properties of Inheritance
Notation

An inheritance relation between two constructions C_1 and C_2 such that C_2 inherits from C_1 will be represented as follows:

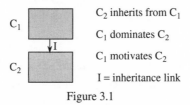

C_2 inherits from C_1

C_1 dominates C_2

C_1 motivates C_2

I = inheritance link

Figure 3.1

inherited information will be represented in *italics;* that is, all information which is shared between the dominating and dominated node is italicized in the dominated construction. As before, profiled information is written in boldface.

Multiple Inheritance Is Allowed

In accord with all of the linguistic applications of inheritance cited above, multiple inheritance paths are allowed. That is, inheritance systems may resemble tree diagrams if each child has only one parent, but in the general case they are "tangled" and can be represented as 'Directed Acyclic Graphs' (DAGs). This allows a given construction in the hierarchy to inherit from more than one dominant construction.

Normal Mode Inheritance

Following Flickinger, Pollard and Wasow (1985) the *normal mode* of inheritance is distinguished from the *complete mode*. The normal mode is designed to allow for subregularities and exceptions, and is the only type to be used here. In the normal mode, information is inherited from dominant nodes transitively as long as that information does not conflict with information specified by nodes lower in the inheritance hierarchy. Lakoff (1984), in his analysis of *there*-constructions, refers to this type of inheritance as "inheritance with overrides" (cf. also Zadrozny & Manaster-Ramer 1993). Normal inheritance is simply a way of stating partial generalizations.

The complete mode of inheritance, which is not exploited here, is designed to capture purely taxonomic relations and constraints. In the complete mode, all information specific to every node which directly or indirectly dominates a given node is inherited. Information from one node may not conflict with that of a dominant node without resulting in ill-formedness. This is the type of inheritance normally assumed in unification-based grammars (e.g., Kay 1984; Fillmore & Kay 1993).

Real Copies: Full-Entry Representations

Fahlman (1979) distinguishes *real copying* from *virtual copying* of information. In real copying, dominated constructions contain all the information that the dominating constructions do: each construction is fully specified, but is redundant to the degree that information is inherited from (i.e., shared with) dominating constructions. This is the type of inheritance employed here. Jurafsky (1992) likens this type of inheritance to the "full-entry" theory of redundancy rules, as opposed to the "impoverished-entry" theory (cf. Jackendoff 1975). Thus the inheritance mechanism of our system is not an on-line process, but rather a static relation defined by shared information (cf. Jackendoff 1975; Aronoff 1976; Bresnan 1978, 1982; Hudson 1984; Lakoff 1984; and Pollard & Sag 1987 for related mechanisms).

In virtual copying, on the other hand, dominated constructions are only partially specified: inherited information is only stored with the dominating construction. Under this mechanism, inferences are computed by searching up the inheritance tree to determine the full specifications of a given construction. This type of inheritance is not exploited here.

Allowing each construction to be fully specified would seem to be an inefficient way to store information; however, this inefficiency is not necessary, depending on the particular implementation adopted. A connectionist system can capture the redundancy without inefficiency by allowing inherited information to be shared information; that is, instead of stating the specifications twice, aspects of the patterns that are inherited are shared by two overlapping patterns. Similarly, in a symbolic system, it is possible to avoid fully specifying particular information twice by allowing particular specifications within constructions to have pointers to other information.

3.3.2   Inheritance Links as Objects

So far we have not said how inheritance links make explicit the particular types of relations that may hold among elements of constructions. That is, inheritance links capture the fact that all nonconflicting information between two related constructions is shared. However, we have not said anything about how to distinguish among various different types of inheritance relations.

In order to make explicit the specific ways that constructions may be related, we will adopt another idea from computer science, that of *object-oriented design*.[2] In particular, the inheritance links themselves will be treated as objects in our system (cf. also Wilensky 1991). Like constructions, they are assumed to have internal structure and to be related hierarchically. Links can be of several types, with various subtypes each. This idea is useful because various kinds of relations among constructions recur in the grammar; in order to capture these generalizations, it is useful to be able to explicitly notate inheritance links as being of specific types. Moreover, as discussed below, by treating links as objects we are able to represent the fact that extensions may be created productively.

Four major types of inheritance links are distinguished: polysemy links, metaphorical extension links, subpart links, and instance links.[3]

Polysemy (I_P) Links

Polysemy links capture the nature of the semantic relations between a particular sense of a construction and any extensions from this sense. The syntactic specifications of the central sense are inherited by the extensions; therefore we do not need to state the syntactic realization for each extension—such specifications are inherited from the dominating construction. The same general type of link is posited to capture morphological polysemy.

Each particular extension is related by a particular type of I_P-link. For example, in chapter 2 it was argued that the ditransitive syntactic pattern is associated with a family of related senses, rather than a single abstract sense. The following pattern of polysemy was observed:

1. 'X CAUSES Y to RECEIVE Z' (central sense)
   Example: Joe gave Sally the ball.
2. Conditions of satisfaction imply 'X CAUSES Y to RECEIVE Z'
   Example: Joe promised Bob a car.
3. 'X ENABLES Y to RECEIVE Z'
   Example: Joe permitted Chris an apple.
4. 'X CAUSES Y not to RECEIVE Z'
   Example: Joe refused Bob a cookie.
5. 'X INTENDS to CAUSE Y to RECEIVE Z'
   Example: Joe baked Bob a cake.
6. 'X ACTS to CAUSE Y to RECEIVE Z at some future point in time'
   Example: Joe bequeathed Bob a fortune.

The caused-motion construction has a strikingly similar pattern of polysemy:

1. 'X CAUSES Y to MOVE Z' (central sense)
   Example: Pat pushed the piano into the room.
2. Conditions of satisfaction imply 'X CAUSES Y to MOVE Z'
   Example: Pat ordered him into the room.
3. 'X ENABLES Y to MOVE Z'
   Example: Pat allowed Chris into the room.
4. 'X CAUSES Y not to MOVE FROM Z'
   Example: Pat locked Chris into the room.
5. 'X HELPS Y to MOVE Z'
   Example: Pat assisted Chris into the room.

In both cases, several of the extensions involve the type of family of related causal relations discussed by Talmy (1976, 1985a, 1985b) under the rubric of "force dynamics." In particular, enablement, resistance, and aiding are concepts force-dynamically related to causation, which is a central component of the central senses of the two constructions. Each of these concepts involves two entities which are construed as interacting via transmission of energy either in the same or in opposing directions (cf. also Jackendoff 1990a for discussion).

Extensions 2, 3, and 4 of the two constructions are quite analogous. The particular verbs involved are different, but the relations between the central sense of transfer or caused motion and the entailments of these extensions is the same. Jackendoff's (1990a) analysis of the infinitive (or "equi") pattern indicates that it, too, has a remarkably similar pattern of interpretations.

At the same time, the full patterns of polysemy in the two constructions analyzed above are not identical. For example, while the caused-motion construction can be used to entail 'X HELPS Y to MOVE Z', no such interpretation is possible for the ditransitive construction:

(1)   *She helped him the prize.
      (Intended meaning: She helped him to get the prize.)

Therefore the patterns of polysemy must in general be learned for each individual construction.

Each of the extensions constitutes a minimally different construction, motivated by the central sense; that is, each sense can be represented by a construction that is minimally different from that of the central sense. The semantic relations are captured by particular I_P-links, and all information about syntactic specifications is inherited from the central sense.

For example, the fifth extension of the ditransitive, sometimes called the "benefactive" construction, can be represented thus, with information that is inherited from the central sense italicized:

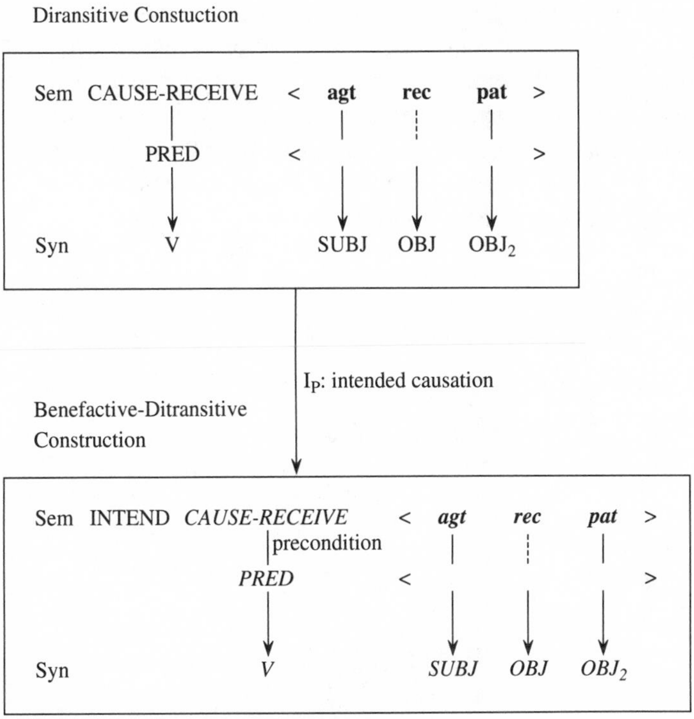

Diransitive Constuction

Benefactive-Ditransitive
Construction

Figure 3.2

The I_P-link between the central sense and the benefactive extension is one that relates causation to intended causation. It licenses expressions such as *Bob baked Mary a cake.*

Since links are objects in the present system, a type of link that recurs often throughout the grammar can be said to have a high type frequency (i.e., there are many instances of the same general type of link) and is therefore predicted to be productively applied to new cases which share the relevant factors associated with the existing cases (cf. chapter 5). Thus, if a polysemy link or any other type of link occurs frequently between distinct constructions with a shared set of characteristics, then that link will be applied to newly learned constructions as a productive form of extension. In this sense, a highly recurrent motivation link is quite analogous to a rule: the existence of one construction will predict the existence of an extension related by the productive link.

Subpart (I_S) Links

A *subpart link* is posited when one construction is a *proper subpart* of
another construction and exists independently. For example, the intransitive
motion construction is related to the caused-motion construction by a subpart
link. The syntactic and semantic specifications of the intransitive motion construction are a subpart of the syntactic and semantic specifications of the
caused-motion construction. The link relating the resultative and intransitive
resultative construction is also a subpart link, since the transitive construction
and intransitive construction are related in exactly the same way here. This type
of relation is diagrammed as follows:

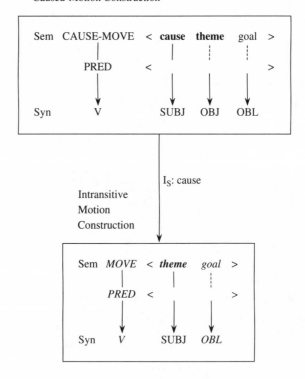

Caused-Motion Construction

Resultative Construction

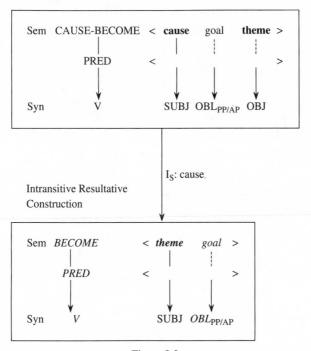

Figure 3.3

Instance (I₁) Links

Instance links are posited when a particular construction is a *special case* of another construction; that is, an instance link exists between constructions iff one construction is a more fully specified version of the other. Particular lexical items which only occur in a particular construction are instances of that construction, since they lexically inherit the syntax and semantics associated with the construction. These cases are therefore treated as partially lexically filled instances of the construction. For example, there is a special sense of *drive* which only occurs in the resultative construction. This sense of *drive* constrains the result-goal argument to mean 'crazy':[4]

(2)   a.   Chris drove Pat mad/bonkers/bananas/crazy/over the edge.
      b.   *Chris drove Pat silly/dead/angry/happy/sick.

The relationship between this sense of *drive* and the resultative construction is represented as follows:

Resultative Construction

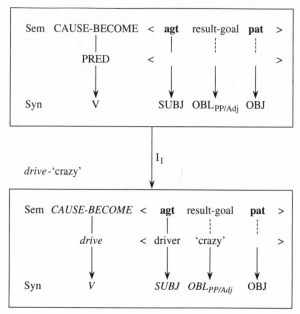

drive-'crazy'

Figure 3.4

Drive's semantics is categorized to be an instance of the CAUSE-BECOME semantics of the resultative construction. Again, all inherited information is represented by *italics.*

Because of the way instance and subpart links are defined, every construction C_1 which is an instance of another construction C_2, and thus *is dominated
by* C_2 via an instance link, simultaneously *dominates* C_2 by a subpart link. That
is, the resultative construction is a subpart of the *drive* lexical construction:

Resultative

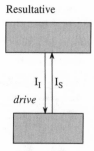

I_I I_S

drive

Figure 3.5

This entails that instances of a particular construction and the construction itself mutually motivate each other.[5] This makes sense insofar as a productive construction is easier to learn given the existence of several instances, while at the same time, conventionalized instances are more likely to exist given the existence of a productive construction.

Because an instance link always entails an inverse subpart link, only instance links will be represented in the diagrams that follow.

Metaphorical Extension (I_M) links

When two constructions are found to be related by a metaphorical mapping, a *metaphorical extension link* is posited between them. This type of link makes explicit the nature of the mapping. The way the dominating construction's semantics is mapped to the dominated construction's semantics is specified by the metaphor. By treating links as objects, it is possible to capture relations among systematic metaphors and ultimately relate the metaphors via an inheritance hierarchy (cf. Lakoff 1993), quite analogous to the hierarchy of constructions. A case of metaphorical extension is discussed in the following section.

3.4   RELATING PARTICULAR CONSTRUCTIONS
3.4.1 The Caused-Motion and Resultative Constructions
A Metaphorical Analysis

In Goldberg (1991b), it is argued that the resultative construction crucially involves a metaphorical interpretation of the result phrase as metaphorical type of goal. Therefore the resultative construction itself, exemplified in (3), can be seen to be a metaphorical extension of the caused-motion construction, exemplified in (4), which involves literal caused motion (see chapter 7 for discussion).

(3)   Pat hammered the metal flat.

(4)   Pat threw the metal off the table.

The idea that these two constructions are related is not new; they are often assumed to be instances of a single more abstract construction.[6] The arguments from Goldberg (1991b) for a metaphorical analysis are summarized below.

The metaphorical analysis allows a wide variety of co-occurrence restrictions to be accounted for. For example, resultatives cannot occur with directional phrases regardless of sequence:

(5)   a. *Sam kicked Bill black and blue out of the room.
          *Sam kicked Bill out of the room black and blue.

b. *Sam tickled Chris silly off her chair.
   *Sam tickled Chris off her chair silly.

At the same time, resultatives *can* occur with prepositional complements that are not directionals:

(6)   a.   Lou talked himself blue in the face about his latest adventure.
      b.   Joe loaded the wagon full with hay.
      c.   He pried the door open with a screwdriver.

Another constraint on the occurrence of resultatives is that they cannot be applied to the theme argument of ditransitive expressions, as (7) shows:

(7)   *Joe kicked Bob a suitcase open.
      (Intended meaning: Joe kicked the suitcase to Bob, causing the suitcase to fly open.)

Also, two distinct resultative phrases cannot co-occur:

(8)   a. *She kicked him bloody dead.
      b. *He wiped the table dry clean.

Finally, as Simpson (1983) and Rappaport and Levin (1991) have pointed out, resultatives cannot occur with directed-motion verbs when used literally. For example:

(9)   a. *She ascended sick.
           (Intended meaning: The ascension made her sick.)
      b. *Jill took the child ill.
           (Intended meaning: The child became ill because of the traveling.)

All of the above restrictions can be explained in the same way as the following, more straightforward example:

(10) *Shirley sailed into the kitchen into the garden.

We need only recognize one fundamental constraint:

> *Unique Path (UP) Constraint:* If an argument X refers to a physical object, then no more than one distinct path can be predicated of X within a single clause. The notion of a single path entails two things: (1) X cannot be predicated to move to two distinct locations at any given time t, and (2) the motion must trace a path within a single landscape.

In the case of literal motion of an object, this constraint is unremarkable. However, the UP Constraint applies not only to literal motion but to metaphorical motion as well.

The stipulation that the motion must occur within a single landscape is meant, then, to rule out examples which would combine literal and metaphorical motion, such as the following:

(11) *The vegetables went from crunchy into the soup.

The relevance of the UP constraint to resultatives becomes clear if resultatives are understood as coding a metaphorical change of location. The necessary metaphor is a general systematic metaphor that involves understanding a change of state in terms of movement to a new location. The mapping involved is simply this:

motion $\longrightarrow$ change
location $\longrightarrow$ state

English expressions reflecting this metaphor include:

(12) a. The jello *went from* liquid *to* solid in a matter of minutes.
      b. He *couldn't manage to pull himself out of* his miserable state.
      c. No one could help her as she *slid into* madness.

By allowing that resultatives code a metaphorical change of location, and understanding the UP Constraint to apply to metaphorical changes of location as well as literal ones, we can explain the co-occurrence restrictions described above. For instance, a resultative is now restricted from occurring with a directional because the directional, coding a change of physical location, will code a path distinct from that coded by the change-of-state resultative. The argument in question is prevented from being understood to simultaneously move to two distinct locations.

Similarly, on this view, the fact that resultatives cannot occur with *arrive, ascend, bring,* and other verbs which imply a physical path stems from the fact that the change-of-state resultative would code an additional, distinct path that would also be predicated of the theme argument.

At the same time, many verbs of directed motion can be used metaphorically to code changes of state. This fact in itself is motivated by the existence of the path metaphor. When used in this way, verbs of directed motion do not necessarily code a distinct path and, as we would expect, can occur felicitously with resultatives as long as a single path is designated. Consider (13a, b):

(13) a. Bob fell asleep.
     b. Bob went crazy.

Example (13a) doesn't mean that Bob literally fell anywhere, but that he metaphorically fell into sleep. Similarly, (13b) doesn't mean that he literally went
anywhere, but that he metaphorically moved to a state of insanity.

To summarize, we can account for the fact that resultatives cannot occur with
directionals, that two resultatives cannot co-occur, that resultatives cannot occur with ditransitives, and that resultatives cannot occur with verbs of motion
when used literally, but can occur with motion verbs when those verbs are used
to imply a change of state, by postulating that the resultative is a metaphorically
interpreted goal phrase.

A metaphorical account of resultatives allows us to explain the lack of polysemy of this construction—the fact that resultatives do not allow the range
of extensions exhibited by the caused-motion construction (or the ditransitive
construction). For instance, resultatives cannot be used to imply an intended,
or potential, change of state:

(14) a. *She allowed it green.
          (Meaning: She enabled it to become green.)
     b. *She locked him dead.
          (Meaning: She prevented him from becoming dead.)

Caused-motion expressions do have these extensions:

(15) a. She allowed him into the room.
          (Meaning: She enabled him to move into the room.)
     b. She locked him out of the room.
          (Meaning: She prevented him from moving into the room.)

This is expected on a metaphorical account since, as was discussed above, metaphorical extensions have as their source domain the *central sense* of the construction. The resultative construction is a metaphorical extension of the central
sense of the caused-motion construction, which is associated with the semantics 'X CAUSES Y to MOVE Z'. Given the metaphorical connection between
movement and change of state, resultative expressions entail 'X CAUSES Y to
BECOME Z', and not 'X ENABLES Y to BECOME Z' or 'X INTENDS to CAUSE Y
to BECOME Z'.

Alternative Analyses

It may be suggested that we can avoid appealing to any metaphorical
interpretation of resultatives by reformulating the UP Constraint as a target

domain constraint. That is, it may be suggested that the constraint be stated as follows:

> *Unique Change of State Constraint:* If an argument *X* refers to a physical object, then no more than one distinct change of state can be predicated simultaneously of *X* within a single clause. Specifically, this means that (1) *X* cannot be predicated to undergo two distinct changes of state at any given time *t,* and (2) any sequence of changes must be understood to involve the same type of change.

This constraint is relevantly similar to that proposed by Levin and Rappaport Hovav (1990a), who follow Tenny (1987) in arguing that resultatives act as delimiters or bounders of events, and that a clause can only be delimited once.

In order for this formulation to account for the co-occurrence restrictions between resultatives and directionals, it is necessary that we consider changes of location to be instances of changing state. What had up to now been analyzed as involving two distinct paths would be reanalyzed as involving two distinct changes of state. In this way, we could try to account for the data cited above without recourse to metaphors.

However, there is reason to prefer the Unique Path formulation to the Unique Change of State formulation. In order for the latter to be viable, we would need to consider *all* changes of location as instances of changing state, not only those which specify a final destination. For example, in (16) the direct object, Bob, would necessarily be understood to undergo a change of state:

(16)   Joe moved Bob toward the door.

But if we generalize the notion of "change of state" to this degree, it seems that undergoing any kind of effect would entail a change of state. This would mean that Bob also undergoes a change of state in (17), for example:

(17)   Joe kicked Bob.

Consequently the proposed Unique Change of State Constraint would be violated by sentences such as (18):

(18)   Joe kicked Bob into the room.

Moreover, it has not been argued that *all* of even the clear instances of changes of state involve the change-of-state metaphor. There is no evidence I know of that simple causative verbs involve this metaphor. For example, although *break* is a causative verb, we have no reason to think that it is necessarily understood in terms of 'X causes y to move to a broken state'. And if we let the UP Constraint be our guide, then there is good reason to think that in fact it does not involve the metaphor. In particular, we find that *break* can occur with a literal directional:

(19)   He broke the walnuts into the bowl.

Further militating against Levin and Rappaport Hovav's (1990a) and Tenny's (1987) formulation of the constraint as one against multiple delimiters is the fact that directionals do not always serve to delimit the event. Directionals can be used to specify a direction without implying any endpoint or delimiting point, as in (20):

(20)   She kicked him toward the door.

However, these nondelimiting directionals are also restricted from occurring with resultatives:

(21) *Sam kicked Bill black and blue toward the door.

Presumably we would like to have the same constraint account for both (5a), *Sam kicked Bill black and blue out of the room,* and (21).

For these reasons, I have chosen to adopt the Unique Path Constraint rather than the Unique Change of State Constraint.

The account presented here of the co-occurrence restrictions described above can also be contrasted with an account proposed by Simpson (1983). She suggests that the co-occurrence restrictions against resultatives occurring with directionals are accounted for by the principle that only one Xcomp—one predicative complement—can appear in a given clause. Her account takes both resultatives and prepositional directionals to be Xcomps. In the case of prepositional directionals, this is a move away from their more traditional categorization as OBL, but it is a reasonable move since directionals can be understood to predicate the theme argument. By distinguishing directionals from other prepositional complements, Simpson's account can satisfactorily explain why resultatives can occur with other prepositional complements but specifically not with directionals. At the same time, depictive predicates are analyzed as Xadjuncts, so they are not subject to the same constraint.

However, Simpson's account fails to predict the fact that resultatives cannot occur with ditransitive expressions. That is, ditransitive expressions are analyzed as involving a Subj, an Obj, and an Obj_θ; the fact that the resultative Xcomp cannot be added is not explained. Moreover, this analysis does not account for why directed-motion verbs, when used literally, cannot occur with resultatives, but *can* when used metaphorically to code a change of state. Finally, this account leaves unexplained why it is that two directionals can co-occur as long as a single path is designated. For example, consider (22):

(22)   Ken drove to L.A. from Pittsburgh.

Notice, we cannot readily claim in this example that a single constituent is involved. What argues against that claim is that *only* can have as its focus anything in its sister constituent (McCawley 1986), yet we find that *only* cannot have as its focus *Pittsburgh* in the following variant:

(23) *Ken drove only [to L.A. from *Pittsburgh*].

This fact argues strongly against treating *to L.A. from Pittsburgh* as a single constituent.

One might think that once we decide that resultatives are a metaphorical extension of the caused-motion construction, nothing more needs to be said. However, there are several reasons for analyzing the resultative and the caused-motion construction as two related but distinct constructions. One reason to keep the constructions distinct is that certain verbs are compatible with only one or the other. For example, *make* only occurs in the resultative construction.[7]

(24)   a.   It made him sick/into a better man.
       b.?*It made him into the room.

Move, on the other hand, cannot occur with the resultative construction:

(25)   a.   He moved it onto the top shelf/away.
       b. *He moved it black.

Moreover, it will be argued in chapter 8 that resultatives can only apply to arguments which potentially undergo a change of state as a result of the action denoted by the verb; that is, resultatives can only apply to arguments which can be categorized as *patient* arguments. This constraint alone serves to distinguish resultatives from caused-motion expressions. Directionals do not require that the argument which they predicate be a patient, only that it be a theme:

(26)   a.   Joe moved it onto the table.
       b.   Joe ran out of the room.

That these arguments are not patients can be demonstrated by their failure to pass Lakoff's (1976) test for patienthood:

(27)   a.??What Joe did to it was he moved it.
       b.?*What happened to Joe was he ran.

In addition, resultatives are subject to several specific constraints that do not hold of caused-motion expressions. For example, it will be argued that resultatives must code an end of scale (section 8.7). The same is not true of directionals:

(28)   a.   He threw it toward the door.
       b.   He put it near the table.

Capturing the Relation

We can represent the relation between the two constructions as follows:

Caused-Motion Constuction

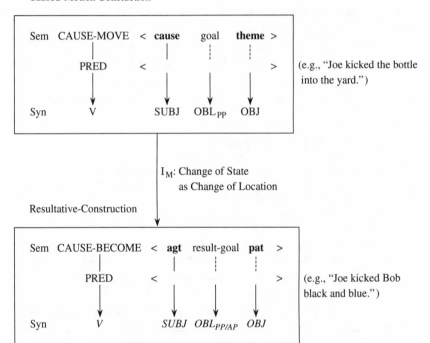

Figure 3.6

As noted above, metaphorical extension inheritance links, or I_M links, are a certain type of inheritance link: the metaphor—in this case, Change of State as

Change of Location—accounts for the relation between the semantics of the two constructions. The syntactic specifications of the metaphorical extension are inherited from the caused-motion construction.

3.4.2   The Ditransitive and Its Prepositional Paraphrase

Many ditransitive expressions can be paraphrased using *to:*

(29) a.   John gave Mary an apple.
    b.   John gave an apple to Mary.

The question that arises, on the account presented here, is not whether verbs are allowed to undergo a lexical or syntactic rule that alters their semantic structure or subcategorization frame, as it is typically taken to be. Rather, the question becomes: How are the semantics of the independent constructions related such that the classes of verbs associated with one overlap with the classes of verbs associated with another? The answer to this question is the subject of this section.

There is a metaphor that involves understanding possession as the "possessed" being located next to the "possessor," transferring an entity to a recipient as causing the entity to move to that recipient, and transferring ownership away from a possessor as taking that entity away from the possessor. Evidence for the existence of this metaphor includes expressions such as the following:

(30) a.   They *took* his house *away* from him.
    b.   He *lost* his house.
    c.   Suddenly several thousand dollars *came into* his possession.

As has previously been suggested by Gruber (1965) and Jackendoff (1972), this metaphor, which we might call "Transfer of Ownership as Physical Transfer," motivates expressions such as the following:

(31)   The judge awarded custody to Bill.
(32)   Bill gave his house to the Moonies.

This metaphor is itself motivated by the fact that giving prototypically correlates with movement from a possessor to a recipient; however, it is clear that such motion is not literally implied by the transfer-of-ownership examples (31–32). Custody does not literally move from the judge to Bill; neither does the house literally move to the Moonies.

The relation between the caused-motion construction and this metaphorical extension can be represented as follows:

Caused-Motion Constuction

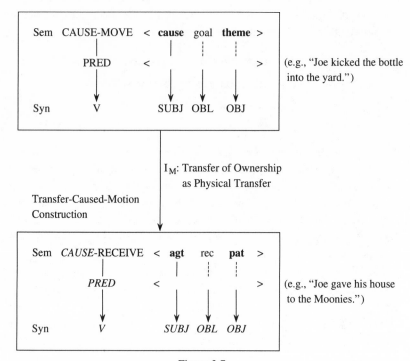

Figure 3.7

The metaphor allows the caused-motion construction to be used to encode the transfer of possession. This is just the semantics associated with the ditransitive construction (cf. chapter 2). We can represent the relation between the caused-motion construction, its extension—labeled Transfer-Caused-Motion Con-struction—and the ditransitive with the following more comprehensive diagram:

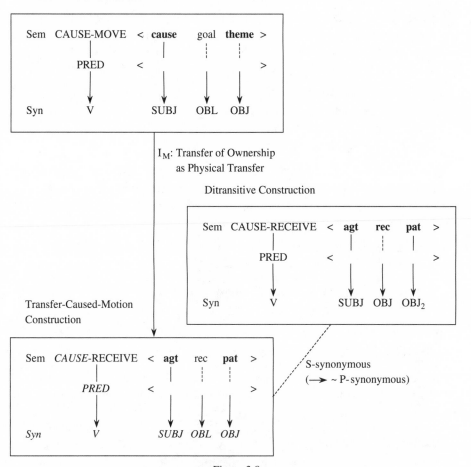

Figure 3.8

The semantic extension (via metaphor) of the caused-motion construction is S(emantically) synonymous with the ditransitive construction.[9] Since the ditransitive construction and the caused-motion construction are not related syntactically, their semantic synonymy (represented by a dashed line in the diagram above) does not constitute a motivation link.

By Corollary A of the Principle of No Synonymy, the two constructions must not be P(ragmatically)-synonymous; that is, the semantic synonymy between them implies a pragmatic difference. And in fact, such a pragmatic difference does exist.

Erteschik-Shir (1979) argues that the ditransitive construction is used when the recipient is nonfocused or "nondominant" (to use Erteschik-Shir's terminology), commonly encoded by a personal pronoun; the transferred object tends to be the focus, and is commonly encoded by an indefinite noun phrase.[10] When these constraints are violated, the expressions are infelicitous:

(33) a. #She gave an old man it.
      b. (She gave it to an old man.)
(34) a. #She sold a slave trader him.
      b. (She sold him to a slave trader.)

On the other hand, the metaphorical extension of the caused-motion construction is used when the focus is on the goal or recipient instead. For instance, example (35) is odd because the transferred object (the house) is the focused information and the recipient is nonfocused.

(35) a. #Mary gave a brand-new house to him.
      b. (Mary gave him a brand-new house.)

Note that examples such as the following are acceptable even though the transferred object is focused and the recipient is nonfocused:

(36) Sally threw a football to him.
(37) Sally handed a scented letter to him.

However, these cases imply physical motion as well as metaphorical motion; that is, they imply that the football or the letter actually moves from Sally to the recipient. Therefore these cases do not require the metaphorical extension of the caused-motion construction but can be based on the literal caused-motion construction itself. Since the ditransitive construction is not S-synonymous with the literal caused-motion construction, no difference in pragmatics is required by the system and no particular pragmatics is claimed here.[11]

We can represent the difference between the ditransitive construction and the metaphorical extension of the caused-motion construction as follows:

Caused-Motion Constuction

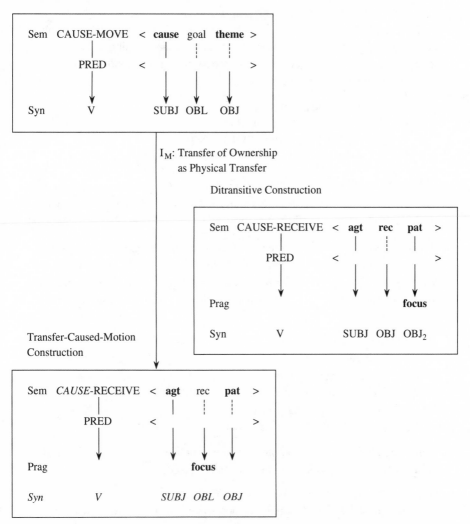

Figure 3.9

The difference in pragmatic structure between the metaphorical extension of the caused-motion construction and the ditransitive construction can be used to explain the puzzle as to why some metaphorical extensions are not felicitous in the prepositional construction. Typically, systemic metaphors whose source domains are compatible with the semantics of a construction license expressions which instantiate the construction. That is, general metaphors may apply to constructions which have the relevant semantics.[12]

However, there are potential metaphorical extensions which do not occur. For example, we have a metaphor that involves understanding an action directed at a person as an object given to the person. This metaphor is evidenced by the following sort of examples:

(38) a. She *threw* him a parting glance.
     b. She *shot* him a keep-quiet look.
     c. She *gave* him a wink/kiss/wave/finger/bow.
     d. She *let* him *have it*.

Notice that this metaphor is somewhat productive:

(39) Bob gave Joe a nudge/a jab/a karate kick to the jaw/a high five/a peck on the cheek.

It is not necessary for a speaker to have heard each of these expressions in order for him to spontaneously generate them or recognize them as acceptable sentences of English. However, we will need to constrain the use of the metaphor to prevent the following (a)-expressions, which involve the caused-motion construction:

(40) a. *She gave a kick to him.
     b. (She gave him a kick.)
(41) a. *She gave a kiss to him.
     b. (She gave him a kiss.)
(42) a. *She threw a parting glance to him.
     b. (She threw him a parting glance.)

Something similar holds for another common metaphor—causation as physical transfer—that is observed in the examples in (43):

(43) a. The idea *presented* her with an opportunity.
     b. The missed ball *handed* his opponent an opportunity on a silver platter.
     c. The noise *gave* me a headache.
     d. The music *lent* the party a festive air.

Again, the following are not acceptable:

(44) a. *She gave a headache to him.
     b. (She gave him a headache.)
(45) a. *The trial gave a lot of grief to her.
     b. (The trial gave her a lot of grief.)

The fact that these metaphorical extensions cannot readily occur with the prepositional construction can be attributed to a difference in their pragmatic specifications. Metaphorical expressions such as *give a kick* focus attention on the action denoted by the nominal, here *a kick*. This is, in fact, what distinguishes *give a kick* from the verbal form *kick,* which can readily be used when the focus is not on the action performed. Similarly, in the metaphorical expressions involving the effecting of some result, the result is typically new or focused information. Therefore the pragmatic properties of the ditransitive argument structure are particularly well suited to expressions such as *give a kick* or *give a headache,* while the pragmatics associated with the caused-motion construction are less well suited.[13] In other words, the metaphorical extension is better motivated as an extension of the ditransitive construction, since as such it can inherit more information; in particular, it can inherit the specification that the action (the metaphorical "transferred thing") and not the recipient is the focused element:

Caused-Motion Constuction

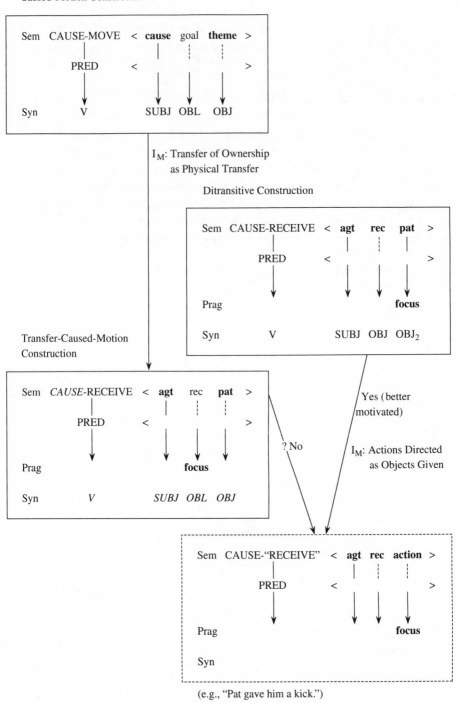

(e.g., "Pat gave him a kick.")

Figure 3.10

Notice that the metaphors can be expressed in the caused-motion construction when the information structure is made more compatible:

(46) When your father comes home, he's really going to give it to you.

(47) Bill gave Mary a kiss and she was so happy that she gave one to everyone she ran into that day.

Thus we find that whether the information structure of a particular metaphorical expression is compatible with the information structure of a particular construction or not plays a role in whether the metaphorical instance is felicitously expressed in that construction. Specifically, the fact that certain metaphorical expressions are not readily expressed in the metaphorical extension of the caused-motion construction, despite the fact that they readily occur in the ditransitive construction, is attributable to the difference in pragmatics between the two constructions.

3.5 MULTIPLE INHERITANCE

Multiple inheritance allows us to capture the fact that instances of some construction types seem to resist being uniquely categorized in a natural way (cf. Borkin 1974; Lakoff 1984). For example, Bolinger (1971) has observed that some instances of the resultative construction pass a test often used as a criterion for the verb–particle construction, in that the resultative phrase can occur either before or after the postverbal NP. He cites examples such as the following:

(48) a. He cut short the speech.
     b. He cut the speech short.

(49) a. Break the cask open.
     b. Break open the cask.

One might be tempted to simply collapse the distinction between resultatives and verb–particle constructions to account for this overlap. However, this move would simply replace one question with another: Why is it that the majority of resultative expressions cannot occur with the resultative phrase placed before the postverbal NP? Consider (50):

(50) a. *He talked hoarse himself.
     b. *He hammered flat the metal.
     c. *He closed shut the door.

Moreover, the verb–particle construction allows particles with an aspectual interpretation, which are not a predicate on the NP argument:

(51) He cleaned the mess up.
      ⇏ The mess is up.

On the present account, examples such as (48–49) above are understood to inherit from two independently existing constructions:

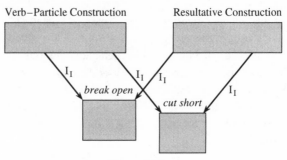

Figure 3.11

Allowing multiple normal mode inheritance links raises the issue of the need for conflict resolution rules. That is, a particular construction may inherit from two other constructions which have conflicting specifications; normal mode inheritance allows for partial inheritance of information when the dominated construction itself has conflicting specifications, but conflict resolution rules would seem to be required for cases in which two or more dominating constructions have conflicting specifications, in order to determine which—if any—of the specifications are inherited.

However, this issue is only relevant if we conceive of inheritance as an online process that is used to predict the specifications of a dominated construction given those of the dominating construction. If instead, as discussed above, each construction is fully specified, any conflict is resolved by an overt specification in the dominated construction.

3.6 INHERITANCE WITHIN CONSTRUCTIONS

Constituents of constructions are also treated as objects in the system—that is, as constructions—and can therefore inherit from other constructions as well (cf. Wilensky 1986). For example, many clausal constructions will have a slot for a OBJ grammatical function. Since we allow the constituents to be constructions themselves, generalizations across constructions as to the semantic and/or syntactic representation of the direct object can be captured by allowing the OBJ function in particular constructions to inherit from a free-standing OBJ construction.

The special sense of *drive* that occurs in the resultative construction inherits from the basic sense of *drive,* since the two senses are related via the metaphor Change of State as Change of Location described above. Thus an I_M-link is posited between them:

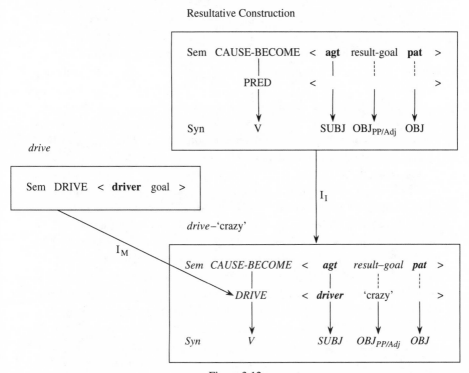

Resultative Construction

Figure 3.12

3.7   CONCLUSION

In this chapter, it has been argued that generalizations about relations among constructions can be captured by conceiving of the entire collection of constructions as forming a lattice, with individual constructions related by specific types of asymmetric normal mode inheritance links. If construction A inherits information from construction B, then B *motivates* A.

Cases of systematically related form and systematically related meaning are formally related by inheritance links. The input and output of some traditional lexical rules are related by inheritance links; an example are the causative and inchoative versions of constructions. In addition, cases not normally related by lexical rule are similarly related by inheritance links. For example, the caused-motion construction is related to the resultative via an I_M-link. In this case, the metaphor Change of State as Change of Location, discussed above, constitutes the systematic relation between form and meaning. In other words, the resultative construction is motivated by the caused-motion construction.

Constructions with unrelated forms, whether or not they encode the same or closely related meanings, are not related by motivation links. For example, *kill* and *die,* although semantically related, are not related directly by an inheritance link. Similarly, the ditransitive construction and its prepositional paraphrase are not related by an inheritance link: neither construction motivates the other.

By allowing inheritance to hold of constituents internal to particular constructions we can capture generalizations about the internal structure of constructions. By allowing multiple inheritance we account for instances which appear to be simultaneously motivated by two distinct constructions.

Moreover, the links themselves are objects in the system, and so they too can inherit from other objects. For example, metaphors which constitute links between constructions can themselves be captured via an inheritance hierarchy (cf. Lakoff 1993). Also, when we treat links as objects, different links can be said to have different type frequencies, depending on how many distinct constructions they relate. A particular link which recurs often throughout the grammar is said to have a high type frequency and, as discussed in chapter 5, is predicted to be productively applied to new cases which share the particular semantic and/or syntactic factors associated with the existing cases. That is, if a link relating pairs of constructions occurs frequently, then that link will be applied to newly learned pairs of constructions which share semantic and/or syntactic characteristics with the existing cases as a productive form of extension. In this sense, a highly recurrent motivation link is analogous to a rule: the existence of one construction will predict the existence of an extension related by the productive link.

Generalizations about the way arguments are mapped onto syntactic structure are discussed in the following chapter.

4 On Linking

In this chapter, the nature of the relation between semantic structure and overt syntactic structure is discussed, and it is argued that many generalizations must be stated at the level of the construction.

In section 4.1, approaches which involve syntactic transformations or derivations are presented—in too broad strokes, no doubt—and the idea of stating linking generalizations in a nontransformational way is defended. In section 4.2, ways to capture generalizations across constructions are discussed.

4.1 TRANSFORMATIONAL APPROACHES TO ARGUMENT STRUCTURE

4.1.1 The Transformational Tradition

In an effort to constrain the application of potentially all-powerful transformations as conceived by Chomsky (1957), Katz and Postal (1964) hypothesized that transformations must be constrained so that they necessarily preserve meaning. Thus the deep structure of a sentence would mirror its semantic structure. Generative Semanticists (e.g., Lakoff 1968, 1972; Langacker 1969; McCawley 1973; Postal 1971; Keenan 1972; Dowty 1972) accepted the hypothesis that transformations preserve meaning and extended it to the position that all and only sentences which are paraphrases of each other should have the same deep structures. Thus upon observing that two sentences bore the same (truth-functional) meaning, researchers set out to derive the two sentences from a single underlying form.

Lakoff (1965) and Partee (1965) presented some of the earliest attempts to derive a systematic symbolic relationship between simple active sentences with systematically related semantics. Lakoff proposed deriving (1b) from (1a) by an optional "flip" transformation:

(1) a. I like it.
     b. It pleases me. (p. 126)

Partee suggested deriving sentences such as (2b) from structures such as (2a) by an optional transformation:

(2) a. John smeared paint on the wall.
     b. John smeared the wall with paint. (p. 85)

She also proposed a transformational relationship between (3a) and (3b):

(3)   a.   John annoyed Mary with his persistence.
       b.   John's persistence annoyed Mary. (p. 36)

Fillmore (1971) proposed deriving caused-motion expressions such as *I hit the ball over the fence* from an underlying structure consisting of two propositions, roughly captured in this case by "My hitting the ball caused it to go over the fence."

Early versions of what was later published as Lakoff (1976) and Lakoff & Ross (1976), giving rise to the tradition of Generative Semantics, proposed that semantic structures actually underlie syntactic structures, and that the base component of grammar generates the set of well-formed semantic structures. A large body of analyses developed this proposal further, essentially arguing that semantic structure needed to be taken into account in order to predict surface structure (e.g. Lakoff 1965, 1968, 1970b; Fillmore 1968; R. Lakoff 1968; McCawley 1968a,b, 1973; Ross 1969, 1970; Postal 1971).

The idea that two forms with the same semantics should be identical at some level of representation has more recently been made explicit in the theory of Relational Grammar, (RG), within which Perlmutter and Postal (1983a) proposed the Universal Alignment Hypothesis (UAH):

> *Universal Alignment Hypothesis:* There exists some set of universal principles which will map the semantic representation of a clause onto the initial grammatical relations.

Although the UAH is not uncontroversial (cf. Rosen 1984), a more specific version of it has been echoed within Government and Binding Theory (GB) in Baker's (1988) Universal Theta Assignment Hypothesis (UTAH):

> *Universal Theta Assignment Hypothesis:* Identical thematic relationships between items are represented by identical structural relationships between those items at the level of D-structure.

These and other more recent proposals are somewhat more constrained than the Generative Semantics analyses were, because transformations or derivations are generally taken to be relevant only in cases of (reasonably) productive morphology (e.g., Baker 1988).[1]

Although these theories differ in many substantive ways, they share the assumption that morphologically related, (truth-functionally) synonymous sentences must share a level of representation. This assumption has guided much of the work in transformational or derivational accounts of argument structure expression. Thus arguments for transformations often run as follows: because the initial level of representation must be the same for all synonymous sentences, if two synonymous expressions are distinct on the surface, one or both must be derived from this shared level of representation.

Generative Semantics and these RG and GB theorists share the important insight that semantics plays a crucial role in determining (underlying) structure. Lakoff states this insight thus: "Syntax and semantics cannot be separated, and the role of transformations and of derivational constraints in general is to relate semantic representations and surface structures" (1971, note 1.65).

4.1.2   Weaknesses of the Transformational Approach

Many of the early proponents of transformations noted semantic differences between forms that were taken to be derived from the same underlying structure (e.g., Partee 1965; Fillmore 1968). They assumed, however, that complete synonymy was not required in order to posit an underlying level of shared representation.

Oftentimes a notion of truth-functional synonymy was and is invoked as the criterion for relating two forms (e.g., Partee 1971). However, with the growing recognition that many aspects of what is intuitively called "meaning" are not captured by truth conditions alone (cf. Fillmore 1975; Fauconnier 1985; Jackendoff 1983; Lakoff 1977, 1987; Langacker 1987a; Pinker 1989; Talmy 1978, 1985a)[2] it is not at all clear why truth-conditional synonymy should have a privileged status in the grammar.

In any case, many expressions commonly related by transformations do not have identical truth conditions. For example, the semantic differences between the following sentences results in a difference in truth conditions:

(4)   a.   He sprayed the wall with paint.
       b.   He sprayed paint onto the wall.

Only in (4a) is it entailed that the wall is somehow "holistically" affected by the paint spraying (Anderson 1971); the most natural interpretation of (4a) is that the wall is covered with paint. By contrast, (4b) would be true if used to refer to a situation in which only a drop of clear paint is sprayed on the wall and the wall is not affected in any way by the paint.

Similarly, the following Chicheŵa examples adapted from Baker (1988) involve a difference in meaning although they are claimed to be related transformationally:

(5)   Mayi   a-nachit-its-a   kuti mtsuko u-gw-e.
       woman 3sg-ps-do-CAUSE-mood that waterpot 3sg-fall-mood
       'The woman made the waterpot fall.'

(6)   Mayi   a-na-gw-ets-a   mtsuko
       woman 3sg-ps-fall-CAUSE-mood waterpot.
       'The woman felled the waterpot.'

As Van Valin (1992) notes, the two sentences are not synonymous: only (6) entails that the causation is "direct." That is, (5) does not entail (6). For instance, Mchombo (personal communication) observes that (5) could be felicitously uttered in a context where a woman was chasing her daughter, and the daughter stumbled over the waterpot as she tried to run away. (6) cannot be felicitously uttered in the same context; the sentence requires that the woman actually makes physical contact with the waterpot.[3]

Semantic differences, when acknowledged, have been accounted for by positing semantic constraints on the application of transformations. However, while adding semantic constraints to syntactic transformations can capture semantic differences, the motivation for postulating syntactic transformations in the first place is often undermined by the existence of these semantic differences. That is, without the assumption of semantic synonymy, many arguments for a derivational relationship lose their force.

This fact was recognized as a weakness of Generative Semantics analyses. Ultimately, the Generative Semantics framework died out in part because of the recognition that rough synonymy was not enough to justify a transformational relationship. In addition, aspects of surface form were shown to be necessary for semantic interpretation (Bresnan 1969; Jackendoff 1969, 1972; Chomsky 1970).

However, the underlying assumption that two forms that are related semantically are necessarily derived from the same underlying form is still implicitly adopted by many theorists (for discussion, see Jackendoff 1990b; Van Valin 1992). To take a fairly recent example, consider Dryer's (1986) in-depth argument for an Antidative analysis. Dryer proposes that ditransitives that can be paraphrased with *to* are in fact more basic than their prepositional paraphrases, and that the latter are derived from the former. For example, (7b) is taken to be derived from (7a):

(7)   a.   Bob gave Sam an apple. (base generated)
       b.   Bob gave an apple to Sam. (derived)

A major aim of Dryer's argumentation is to defend the existence of two grammatical relations, Primary Object (PO) and Secondary Object (SO). The PO corresponds to the direct object of transitive clauses and the first object of ditransitive clauses; the SO corresponds specifically to the second object of ditransitive clauses. The central argument for the existence of these categories is based on an appeal to what Dryer calls the "Natural Class Principle," which states that if many languages have rules that apply specifically to a certain form, then this form should be treated as a natural class.[4] Since passive and object marking are sensitive to the PO in many languages, the Natural Class Principle implies that the PO exists as a natural class. Further evidence for the existence of distinct categories PO and SO comes from word order and case marking facts. Let us assume the following (partial) ordering, which takes the distinction between PO and SO into account:

(8)   Subj > DO
      Abs > Erg
      DO > IO
      PO > SO[5]

The principles governing English word order and case marking facts can now be stated quite simply: All and only terms (i.e., Subj, DO, IO, PO, SO) are not marked with prepositions; the word order must be: Subj-V-PO/DO-SO/IO-nonterms.

Dryer goes on to argue that this fact—the simplification of word order and case marking descriptions given the existence of the categories PO and SO—is evidence for his Antidative analysis. But in order for this fact to substantiate his claim that prepositional paraphrases with *to* are derived from ditransitive expressions. Dryer must assume (1) that the two forms necessarily share a level of representation, and (2) that derivations cannot create grammatical relations (therefore, if PO and SO exist, they must be base generated). If we do not assume (1), with its implication that either the ditransitive or the prepositional paraphrase is derived, there is no reason why the two forms cannot both be base generated. That is, the argument as to whether the categories PO and SO exist in English has no bearing on the question whether an alternation account is warranted or not; Dryer provides no independent evidence for the derivational analysis.[6]

Not all transformational or derivational accounts rely crucially on an underlying shared representation between pairs of expressions that share a rough semantic equivalence (e.g., Perlmutter 1978; Perlmutter & Postal 1983b; Aissen 1983; Farrell 1991). However, there are other reasons to avoid a transformational relationship between related constructions if possible.

Bowerman (1982) and Gropen et al. (1989) show that in child language acquisition, semantic restrictions are operative as soon as certain constructions are produced, there being no period of unconstrained overgeneralization on the basis of a purely syntactic relation.[7] For example, Gropen et al. (1989) show that the semantic restriction that the recipient of a ditransitive must be animate is operative as soon as the ditransitive syntax is produced. Thus none of the following possible types of overextensions were ever uttered by any of the children they observed:

(9) *Amy took Chicago Interstate 94. (Amy took Interstate 94 to Chicago.)

(10) *Betty threw the tree the box. (Betty threw the box to the tree.)

(11) *Alex put his head a gun. (Alex put a gun to his head.)

(12) *Babs took fun a trip. (Babs took a trip for fun.) (Gropen et al. 1989:218)

As Gropen et al. note, this calls into question the idea that the dative rule is fundamentally a syntactic operation; there is no clear reason why a syntactic operation would be instantaneously constrained by an arbitrary semantic condition. Moreover, since an unconstrained rule would be easier to learn and represent and would provide more expressive power (Pinker 1989), it is not clear why the semantic constraint on this putative syntactic rule is not ignored by new generations of speakers.

Another problem with approaches that rely on transformations is that they posit an often unwarranted asymmetry between two constructions that are thought to be related. In the case of the ditransitive, *He gave the book to her* is usually supposed to be more basic than *He gave her the book* (contra Dryer 1986). A typical reason given is that the verbs which allow ditransitives are a proper subset of those that allow prepositional paraphrases. However, this is not actually so: *refuse* and *deny* do not have paraphrases with *to* or *for,* and neither do many metaphorical expressions. For example:

(13) a. She gave me a headache/a kiss/an idea.
      b. *She gave a headache/a kiss/an idea to me.

Moreover, Oehrle (1976) has argued that there is no principled way to distinguish those cases which have prepositional paraphrases from those that do not.

Developmental data (Gropen et al. 1989) shows that the ditransitive and prepositional paraphrases occur at roughly the same time in children's speech, with neither construction reliably preceding the other, so that evidence for an asymmetry cannot be grounded in evidence from children's acquisition of the forms.

Consider also the English locative alternation. In general, the pattern associated with (14) is supposed to be more basic than that associated with (15) (cf. Channon 1980; Perlmutter & Postal 1983a).

(14) He loaded hay onto the wagon.

(15) He loaded the wagon with hay.

However, when different verbs are examined, this claim of asymmetry is not clearly warranted. So, although *stack* and *plaster* allow both argument structures, there is no intuition that the *onto* variant is more basic than the *with* variant. That is, the following appear to have equal status in terms of being basic or unmarked:

(16) a. He stacked the shelves with boxes.
      b. He sacked boxes onto the shelves.
(17) a. He plastered the wall wtih posters.
      b. He plastered posters onto the wall.

Moreover, *adorn, blanket, block, cover, dam, enrich, fill, dirty, litter, smother, soil, trim, endow, garnish, imbue, pave, riddle, saturate*—to name a few—only occur with the *with* variant. In fact, in a detailed study of locative verbs, Rappaport and Levin (1985) found that out of 142 verbs studied, only 34 alternated, with exceptions existing in both directions (Pinker 1989).

On a constructional approach, we need not assume an asymmetrical relationship between two constructions that are found to be related. We can describe instances of partial overlap of syntax, semantics, or pragmatics as such, without necessarily assuming that one of the constructions involved is basic, the other derived. For example, we can state that the semantics associated with the ditransitive construction is related to the semantics of the paraphrase with *to;* we do not need to assume the primacy of one over the other. And we can describe similarly the relations between paraphrases with *to* and other instances of the caused-motion construction, for instance between the following (a) and (b) expressions:

(18) a. Ethel brought the wrench to Fred.
      b. Ethel brought the wrench toward Fred.
(19) a. Ethel threw the ball to Harry.
      b. Ethel threw the ball over Harry.

These cases are discussed in chapter 7.

To summarize, accounts of argument structure which relate one construction to another by a syntactic transformation that derives one from the other have several drawbacks:

  1. Expressions that are claimed to share a level of representation are not fully synonymous. This raises the following problems:

(a) Which aspects of semantics are relevant to determining semantic equivalence, and thus a shared level of representation, has never been adequately detailed.

(b) In many cases the only motivation (often implicit) for proposing a derivational relationship in the first place is semantic synonymy.

2. The semantic distinctions are learned as early as the forms themselves, which casts doubt on the idea that the transformations are basically or primarily syntactic.

3. Such accounts postulate an asymmetry between the two forms in question. However:

(a) There are typically lexical items that only have the output form of a putatively optional transformation.

(b) The two forms are often learned at roughly the same age (with neither one reliably preceding the other).

While transformational accounts explicitly represent semantic relations among constructions, the constructional approach, as we saw in the previous chapter, takes a different view. On the constructional approach, semantic similarities that do not coincide with formal similarities are captured implicitly, because of a relation between the specified semantics, but are not explicitly notated in the grammar. The intuition is that the existence of a given form with a particular meaning in no way motivates the existence of a different form with a closely related meaning. Therefore, inheritance links are not posited between constructions that are not related formally. Only relations involving both form and meaning (or sometimes just form; cf. Fillmore & Kay 1993) are explicitly represented by positing inheritance links.

An apparent benefit to transformational approaches is that they allow the relationship between underlying form and meaning to be stated in a straightforward, and often transparent, way.[8] The question arises, how are cross-linguistic generalizations about the relationship of semantic representation to *overt* syntactic expression to be captured within a constructional approach? This is the subject of the following section.

4.2   GENERALIZATIONS ACROSS CONSTRUCTIONS

Generalizations across constructions concerning word order facts, case-marking properties, and links between semantics and grammatical relations can all be captured by stating these generalizations at a sufficiently high node in an inheritance hierarchy of constructions. Thus, such generalizations are inherited through dominated constructions, unless a particular construction prevents such inheritance by having a conflicting specification. The following is an overview of all the relations discussed in the previous chapter:

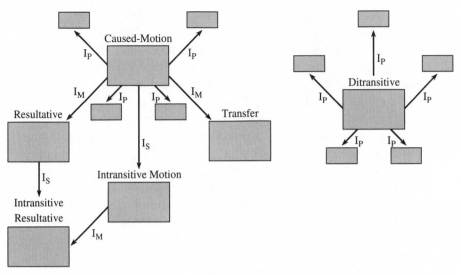

Figure 4.1

Leaving the polysemy and metaphorical extensions out of the diagram, we can represent some of the more general relations among constructions as follows:

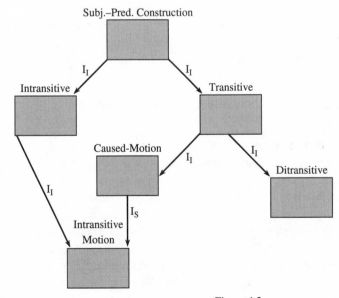

Figure 4.2

The fact that English is an SVO language can be captured by specifying a word order constraint on the top node of the diagram, at the level of the subject–predicate construction. Certain constructions further down the inheritance hierarchy, such as the topicalization construction or the locative *there* construction (not shown), can override the word order constraint with construction-specific constraints. Thus generalizations about word order can be captured while at the same time other constructions with exceptional word orders are permitted. Subregularities are expressed similarly by stating a generalization at a node that is intermediate on the hierarchy.

It should be stressed that if a generalization is construction-specific this does not entail that it is not part of a recurring pattern crosslinguistically. We know that many languages have constructions closely analogous to, for instance, the English transitive, ditransitive, locative, and topicalization constructions. It is quite possible that there is a universal inventory of possible argument structure constructions relating form and meaning, and that particular languages make use of a particular subset of this inventory.

Along with many other theories of thematic roles, ours makes no assumption that thematic roles are primitives (cf. Jackendoff 1972, 1983, 1987; Foley & Van Valin 1984; Rappaport and Levin 1988; Gropen et al. 1989; Pinker 1989; Van Valin 1990b). Instead, roles are taken to be slots in relational semantic structures. *Argument roles* are defined to be slots in the semantic representation of particular constructions and *participant roles* are defined to be slots in the rich semantic representation of predicates (cf. chapter 2). The linking of semantics to syntactic expression is claimed to be generally determined within constructions, that is, at the level of argument roles. At the same time, exceptional linking patterns may be stated as part of particular lexical entries (cf. Fillmore & Kay 1993).

4.2.1 Empirical Weaknesses of Construction-Independent Linking Rules

It might be tempting to think that individual constructions are not the right level at which to capture generalizations about syntactic expression, and that instead very general linking rules mapping particular roles onto grammatical relations or syntactic configurations should be a priori preferable.

Such general linking theories have been proposed for some time. For example, Fillmore (1968) suggested that subject selection was determined with reference to a fixed thematic role hierarchy; the highest available role on the hierarchy would be mapped onto the subject. More recent attempts to relate argument structure and overt syntactic form in a general way can be found in, for example, Foley & Van Valin 1984, Carter 1988, Pinker 1989, and Rappaport & Levin 1988. Such linking theories are motivated by the fact that there are intra- and inter-language generalizations about the kinds of complements particular predicates have. The attempt, then, is based on the fact that clearly, syntactic form is not related in an arbitrary way to the semantics of predicates.

In this section, I first review evidence that in a monostratal account, construction-specific linking rules are required—that it is not possible to state all linking generalizations in a construction-independent way: certain mappings of semantic arguments to grammatical forms are only relevant to particular constructions (cf. also Koenig 1993). The two argument roles discussed are those of *recipient* and *theme*.

Recipients

In English, recipient arguments (or the first argument of an abstract predicate HAVE or RECEIVE) can be linked to three different grammatical relations. Which grammatical relation is actually expressed depends on the construction at hand. For example, in the ditransitive construction recipient arguments are expressed as objects:

(20)   Sam gave *Mary* a cake.
      Subj   V   *Obj*   Obj$_2$

They also appear in oblique phrases in the transfer-caused motion construction:[9]

(21)   Sam gave the piece of land *to his son*.
      Subj   V   Obj   *Obl*

Recipient arguments are also sometimes expressed as subjects:

(22)   *Sam* received/got/acquired a package.
      Subj   V   Obj$_?$

What we have then is the situation diagrammed below. The same generally defined argument occurs overtly in different syntactic positions (bearing different grammatical relations). Which syntactic position is actualized is determined by the construction, not by the thematic role in isolation.

Recipient

Obj$_1$ Subj Obl

Figure 4.3

Thus the mapping from semantics to grammatical relations is not determined by a function that is based solely on the thematic role to be expressed. Instead, we find cases wherein the syntactic expression is construction specific.

Themes

Consider what the "theme" argument would be mapped onto in a construction-independent account if the "theme" is defined to be an argument which undergoes a change of state or location. A specific attempt at such an argument has been formulated within the monostratal linking theory of LFG by L. Levin (1987), Alsina and Mchombo (1990), Bresnan and Kanerva (1989), Bresnan and Moshi (1989), Bresnan and Zaenen (1990), and Ackerman (1990). This theory is chosen here for discussion because to my knowledge it is the most detailed attempt at a linking theory within a monostratal framework. In this theory, grammatical relations are predicted from the argument structure of particular predicates. Argument structures are represented by argument (theta) role arrays, although there is no strong assumption that the argument roles are primitives instead of being derived from a richer decompositional semantics. In fact, most proponents of this theory suppose that the roles are shorthand for different argument places in some logical decomposition in the style made familiar by Generative Semantics (see Jackendoff 1972, 1983, 1987; Foley & Van Valin 1984; Rappaport and Levin 1988; Gropen et al. 1989; and Pinker 1989 for arguments that thematic roles are not primitive).

Two abstract features are postulated, [r] and [o], which categorize four types of grammatical relations:

$$\text{SUBJ } [-r, -o] \quad \text{OBL}\theta \ [+r, -o]$$
$$\text{OBJ } \ [-r, +o] \quad \text{OBJ}\theta \ [+r, +o]$$

(OBJθ denotes the second object of ditransitives.) The feature [r] stands for (*semantically*) *restricted;* [o] stands for *objective* or *object-like.* Thematic roles are assigned features in two ways. On the one hand, they have an intrinsic classification (IC), which is said to be based on their inherent semantic properties. A first approximation of this basic classification is given below.

Intrinsic Classification (first approximation):
- theme roles: $[-r]$
- all other roles: $[-o]$

On the other hand, roles receive a default assignment: the highest theta role on the proposed hierarchy receives a $[-r]$ feature as a default, the rest receive $[+r]$ (Bresnan & Moshi 1989; Alsina & Mchombo 1990). The hierarchy that is adopted is:

agent > beneficiary > goal > instrument > theme > location

Alsina and Mchombo (1990) propose that applicatives in Chicheŵa are formed by a lexical rule which adds a "dependent" argument to the argument structure of the matrix verb; (23) illustrates this for *cook:*[10]

(23) $cook_0$ <agt pat> $\Rightarrow$ $cook_1$ <agt $\theta_{dependent}$ pat>

The "dependent"-subscript on the theta role 'θ' is intended to capture a semantic property that is claimed to be loosely correlated with affectedness. Other LFG accounts have described this semantic property as "applied" (Bresnan & Moshi 1989), "patient-like" (Bresnan 1990), or "affected" (Ackerman 1990). Unfortunately, this attribute is not fully explicated in any of these analyses. For the sake of consistency, I will refer to it as "dependent" throughout.

The rule in (23) is analogous to the semantics-changing lexical rules proposed in Levin & Rapoport 1988 and Pinker 1989. Several difficulties stemming from the semantic claims inherent in this type of approach have been discussed in chapter 1. The primary focus within LFG, however, is on the linking between lexical semantics and surface syntax; it is this aspect of the approach which is considered here.

Alsina and Mchombo (1990) state the intrinsic classification assignment for the applicative construction in general terms:[11]

(24)   When there is a theme and a(nother) dependent argument, then one will receive $[-r]$ and the other will receive $[+o]$ (in languages that allow only one direct object, like Chicheŵa and English).

Dependent *recipient* roles have a special status, since when present, they must occur directly after the verb, can be expressed as an object-marker on the verb, and can be passivized, whereas the co-occurring theme argument cannot (Morolong & Hyman 1977; Alsina & Mchombo 1990). In these ways the recipient dependent argument is direct-object-like, and like other direct objects may appear as subject in passives. These facts are accounted for if a dependent recipient is necessarily either a surface SUBJ or OBJ. Alsina and Mchombo therefore propose that dependent recipient roles must receive a $[-r]$ classification, which distinguishes SUBJ and OBJ relations from other relations.

We thus have the following revised list of intrinsic classifications now:

Intrinsic Classification (revised):
- theme/patient roles: $[-r]$ or $[+o]$
- "dependent" recipient role: $[-r]$
- other dependent roles: $[-r]$ or $[+o]$
- all other roles: $[-o]$

Another example of a semantics-changing lexical rule is proposed by Ackerman (1990) in order to account for the locative alternation involving verbs such as "spray" and "load" in Hungarian (examples are Ackerman's):

(25)　a paraszt　　(rá=)rakta　　　　　　a szénát　　a szekérre
　　　the peasant (onto)loaded-3sg/DEF the hay-ACC the wagon-SUBL
　　　'The peasant loaded$_0$ the hay onto the wagon.'

(26)　a paraszt　　meg=rakta　　　　　　a szekeret　　szénával
　　　the peasant PERF-loaded-3sg/DEF the wagon-ACC hay-INSTR
　　　'The peasant loaded$_1$ the wagon with hay.'

The locative argument of *meg=rakta* in (26) is a dependent argument, which gets the intrinsic classification [− r] just as dependent arguments do on Alsina and Mchombo's account. Thus we have the following:

(27)　"load$_0$" <agt loc theme> ⇒ "load$_1$" <agt loc$_{dependent}$ theme>

The *with* variant given in (26) seems to fit squarely within the domain of generalization in (24), which would predict that the theme argument should receive a [+o] intrinsic classification, ultimately resulting in its being linked to an OBJ$_\theta$ relation like the theme argument in applicative constructions. However, as (26) shows, the theme argument in the presence of another dependent argument is not linked to OBJ$_\theta$ but rather mapped to OBL. In order to avoid the conclusion that locatives should be expressed in a way directly parallel to applicatives, Ackerman stipulates that the theme argument does not get the intrinsic classification [+o], but instead receives [+r] as intrinsic classification and then [− o] as a default.[12]

For languages that have ditransitive (or applicative) expressions and both forms of the locative alternation, for example, English and Chicheŵa,[13] we now need to postulate the following intrinsic classifications:

Intrinsic Classification (final formulation):
1. If there is another dependent role which is a recipient, theme/patient role: [+o]
2. If there is another dependent role which is a locative, theme/patient role: [+r]
3. If there is no other dependent role, theme/patient: [− r]
4. dependent recipient role: [− r]
5. all other roles: [− o]

These rules are in part language specific, since not all languages have all of the relevant constructions. More crucially for our purposes, observe the three feature-assigning rules (1 − 3) involving the theme role. Each of these rules is sensitive to other roles present in the argument structure. This context sensitivity is expected on a constructional approach to linking, but may have been precisely what the LFG linking theory had specifically wanted to avoid (Mchombo, personal communication).

Taking the Alsina & Mchombo and Ackerman analyses together, we find that the theme can receive the intrinsic classification [−r], [+o], or [+r] (depending on what other roles are assigned), that is, all but one possible assignment ([−o]). It becomes difficult to see what is supposed to be meant by "intrinsic classification." That is, what is called "intrinsic classification" is assigned not on the basis of intrinsic properties of the argument, but rather on the basis of properties of other, co-occurring arguments. Moreover, as Ackerman (1992, note 21) observes, this assignment of features still does not account for the expression of themes as subjects in the intransitive motion construction.

Adding to this difficulty of too wide a range of possible intrinsic classifications for a given role is the fact that these classifications are not assigned on the basis of independent evidence. What determines which of the possible classifications is actually assigned is what construction is supposed to be predicted. In this way, the classifications become circular: the recipient is dependent, and thus [−r] as opposed to [−o], just in case it is supposed to be the OBJ in the ditransitive construction. Since the notion "dependent" is never adequately defined, no independent criterion for the assignment of dependent status is offered. In short, since the linking theory can *by its nature* capture the necessary facts—because the abstract features [r] and [o] are all that is needed to code the grammatical relations—unless independent criteria for assigning grammatical relation features are found, the formalism only serves to code the syntactic structure that is supposed to be predicted.

The general point is that the linking of the theme role to an overt grammatical relation crucially depends on what other arguments are present. The same point can be made equally well with English data. The theme argument, that is, the entity whose motion or location is at issue, can occur as subject in the intransitive motion construction:

(28) *The boy* ran home.

As object in simple transitives:

(29) Pat moved *the bat.*

As oblique in a version of the locative alternation:

(30) Pat loaded the truck *with hay.*

And as the second object in ditransitives:

(31) Pat threw Chris *the ball.*

Which grammatical relation is actually realized depends on which construction is expressed.

4.2.2   Capturing Linking Generalizations with Constructions

Once we recognize the large number of construction-specific linking rules, the question arises whether there are any more general linking rules. In this section, Dowty's (1991) account of linking generalizations is discussed, and it is shown that his observations can be adapted to the present framework quite naturally.

Dowty (1991), following the spirit of Foley and Van Valin (1984), suggests two general macro-role types: *Proto-Agent* and *Proto-Patient* (cf. Foley & Van Valin's "Actor" and "Undergoer"). These roles are defined as prototype concepts, much like the prototypes discussed by Rosch and her colleagues (e.g., Rosch and Mervis 1975). The lists of Proto-Agent and Proto-Patient properties draw in part on the empirical findings of Keenan (1976, 1984), who detailed many properties associated with Subject and Object crosslinguistically.

Proto-Agent properties:
1. volitional involvement in the event or state
2. sentience (and/or perception)
3. causing an event or change of state in another participant
4. movement (relative to the position of another participant)
5. exists independently of the event named by the verb

Proto-Patient properties:
1. undergoes change of state
2. incremental theme
3. causally affected by another participant
4. stationary relative to movement of another participant
5. does not exist independently of the event, or not at all.

Given these definitions of Proto-Patient and Proto-Agent properties, Dowty proposes the following principle:

> *Argument Selection Principle:* In predicates with grammatical subject and object, the argument for which the predicate entails the greatest number of Proto-Agent properties will be lexicalized as the subject; the argument for which the predicate entails the greatest number of Proto-Patient properties will be lexicalized as the direct object.

Proto-roles reflect higher-order generalizations about lexical meanings. Therefore, the principles above allow for a small number of lexical exceptions (e.g., *undergo, sustain, tolerate, receive, inherit*).[14]

In syntactically ergative languages, for example Dyirbal (Dixon 1972), certain Mayan languages including Mann (England 1983) and Quiche (Treschsel 1982), the subject in transitive clauses is the argument with more Proto-Patient properties. For these languages Dowty reverses the syntactic association. He notes, however, that "what we do not find, even in split ergativity, is 'random' alignment from one verb to another, e.g. 'build' with Agent absolutive but 'kill' with Patient absolutive" (p. 582).

It turns out that Dowty's linking generalizations are naturally accounted for in the present framework. Notice the domain of application of Dowty's principle: "In predicates with grammatical subject and object. . . ." Clearly the principle is only relevant if the transitive construction is involved. Therefore, the generalization that if there is a SUBJ and an OBJ, then the role that is more agent-like, the "proto-agent," is linked with SUBJ and the "proto-patient" role is linked with OBJ, can be captured by specifying these linking rules within a skeletal transitive construction, and allowing other constructions to inherit from this construction.

Transitive Construction

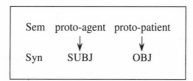

Figure 4.4

In syntactically ergative languages, the transitive construction has the reverse linking, so that SUBJ is linked with the proto-patient role and OBJ is linked with the proto-agent role; these linkings are then inherited by other constructions as long as those constructions' specifications do not conflict.

Just like Dowty's proposal, the constructional account allows for a limited number of lexical exceptions. Exceptions are cases which do not inherit from (i.e., are not motivated by) the transitive construction, or cases which inherit only the form but override the meaning of the construction. The number of exceptions is limited because such non-motivated (or less well motivated) cases exist at a cost to the overall system, in accord with the Principle of Maximized Motivation (cf. chapter 3). Thus the inheritance hierarchy allows us to capture the relevant generalizations while at the same time allowing for a limited number of lexicalized exceptions.

4.2.3   Some Speculations on the Semantics of the Transitive Construction

By being posited as a unitary structure with semantics consisting merely of two abstract "proto-roles," the transitive construction in figure 4.4 is represented as if it had a single, very general abstract meaning. It may well turn out that this construction is more like the others discussed in this monograph in that it may be more felicitous to assign a family of related meanings to it, with the prototypical "transitive scene" (cf. Lakoff 1977; Hopper & Thompson 1980; Rice 1987a,b) being the central sense. In this case, the central sense would be quite specific, being that of a volitional actor affecting an inanimate patient—a causative event.[15] Extensions from the prototype would license a wider range of transitive expressions.

One indication that multiple senses may be involved stems from the fact that there exist clusters of cases which are not instances of the general semantic template given in figure 4.4. For example, alongside the "exceptional" *receive* and *inherit,* we find the closely related verbs *have, own, acquire, get.* The existence of so many cases militates against the idea that they are in fact exceptional. On a multiple-sense view, it is possible to posit additional senses of the transitive construction, for example a sense 'HAVE (X, Y)' (cf. Pinker 1989).

Another benefit of the multiple-sense view is that it is able to capture the generalization, discussed in chapter 2, that crosslinguistically language learners apply transitive markers first to expressions designating prototypical transitive scenes, before extending them to less prototypically transitive expressions (Slobin 1985).

Languages differ in how and to what extent the transitive construction is extended to express nonprototypical semantically transitive scenes. In recognizing distinct senses, we might be able to develop a model in which we would be able to isolate the true crosslinguistic generalizations. In other words, languages differ in how they express noncanonical transitive predicates, such as predicates of possession ("have", "own"), but languages express semantically canonical transitive predicates by means of a transitive construction.[16] Language-specific idiosyncrasies would arise, according to this view, in just how languages extend their inventory of grammatical argument structure constructions to cover expressive requirements.

Linking generalizations apply to the basic senses of constructions. Additional senses of the transitive construction, related to it by polysemy links (cf. chapter 3), would inherit the linking specifications of the construction they are dominated by. Therefore, once we posit extensions of the basic transitive construction, the fact that these extensions have the same syntactic expression would follow.

4.3 CONCLUSION

To summarize, this chapter has presented general arguments for a monostratal approach to the relation between overt syntactic expressions and semantic representations. It has been suggested that the degree of regularity in the relation between semantic role types and overt syntactic expression is sometimes exaggerated, and that many linking generalizations are construction specific. The cross-constructional generalizations that do exist are naturally captured in the present framework by stating the relevant regularities at a high node in the hierarchy of constructions; subregularities or minor patterns are captured by stating the respective generalizations at intermediate nodes. Exceptions are allowed to exist, but only at a cost to the overall system.

5 Partial Productivity

> If you invent a verb, say *greem,* which refers to an intended act of communication by speech and describes the physical characteristics of the act (say a loud, hoarse quality), then you know . . . it will be possible to greem, to greem for someone to get you a glass of water, to greem to your sister about the price of doughnuts, to greem "Ecch" at your enemies, to have your greem frighten the baby, to greem to me that my examples are absurd, and to give a greem when you see the explanation.
>
> Arnold Zwicky (1971)

5.1 INTRODUCTION

It has been a long-standing puzzle that many constructions are used somewhat productively (as implied by the above quotation),[1] yet resist full productivity. This chapter addresses the issue of partial productivity for the most part by examining the ditransitive construction as an example. In section 5.5.2 other constructions, which can be seen to be either more productive or less productive, are considered.

The ditransitive construction can be used somewhat productively; that is, the construction can be extended to new and hypothetical verb forms (e.g., Wasow 1981). For example, the new lexical item *fax* can be used ditransitively as in (1):

(1)   Joe faxed Bob the report.

Also, hypothetical lexical items are readily adapted to the ditransitive syntax. As Marantz (1984) notes, if we define a new verb, *shin,* to mean "to kick with the shin," it is quite natural for us to allow this new verb to be used ditransitively, as in (2):

(2)   Joe shinned his teammate the ball. (p. 177)

Experimental evidence confirms the fact that speakers extend constructional patterns for use with novel verbs (Pinker, Lebeaux & Frost 1987; Pinker 1989; Maratsos et al. 1987; Gropen et al. 1989, 1991; Braine et al. 1990).

At the same time, the ditransitive pattern is not completely productive within any generally defined class of verbs. Seemingly closely related words show distinct differences as to whether they allow ditransitive syntax. The following contrasts exist in many dialects:

(3)   a.   Joe gave the earthquake relief fund $5.
       b. *Joe donated the earthquake relief fund $5.

(4)   a.   Joe told Mary a story.
       b. *Joe whispered Mary a story.

(5)   a.   Joe baked Mary a cake.
       b. *Joe iced Mary a cake.

Brown and Hanlon (1970) have argued that children are neither corrected nor miscomprehended more often when they speak ungrammatically, so that they have no recourse to "negative evidence" that could allow them to either un-learn or avoid learning the above type of ungrammatical sentences (cf. Braine 1971; Baker 1979).

The standard solution to the no-negative-evidence problem in the case of vocabulary learning is to assume that there is indirect negative evidence in the form of attested input, assuming a principle that synonymy is avoided (cf. dis-cussion in chapter 3). That is, a child may overgeneralize the past tense con-struction to produce *comed* as the past tense of *come,* but upon learning that *came* is synonymous, the child will expunge *comed* from her vocabulary, since she will assume that the language does not have two terms *comed* and *came* which are synonymous. Such indirect evidence is not forthcoming in an obvi-ous way in the case of alternative syntactic patterns. It is not likely that the child simply expunges (6) upon hearing (7) because many verbs (e.g., *give*) do occur in both forms.

(6)   *He whispered the woman the news.

(7)   He whispered the news to the woman.

Moreover, as noted above, experimental evidence shows that children do not learn how to use syntactic patterns entirely conservatively, that is to say, solely on the basis of the input. If properly primed, they are willing to extend their use of verbs to previously unheard but related patterns.

An apparent paradox arises then, since if speakers have a productive mecha-nism that allows them to extend the use of the ditransitive syntax to new and novel verbs, it is not clear what prevents them from overgeneralizing to pro-duce the above ill-formed examples (3b–5b).

This paradox is often sidestepped in linguistic theories. Thus, whether relation-changing lexical rules are intended to be purely redundant generalizations over stored items in a fixed lexicon, or rather generative rules which produce new forms productively, is often not made entirely clear.

Jackendoff (1975), for example, states that his lexical rules are intended only to account for existing regularities (both morphological and semantic) within the lexicon. These rules are represented by two-way arrows which encode the symmetric relation "is lexically related to." This aspect of Jackendoff's account is crucial, since he argues explicitly against Lakoff's (1965) proposal that productive rules generate "hypothetical lexical entries." However, Jackendoff also suggests that "after a redundancy rule is learned, it can be used generatively, producing a class of partially specified possible lexical entries" (p. 668).

Bresnan (1982) also attempts to find a middle ground between nonproductive and fully productive rules. While the lexical rules of LFG are explicitly conceived of as "redundancy rules," the metaphor of a lexically changing process is pervasive. The following is Bresnan's early description of the passive lexical rule:

Passivization in English
Functional *change:* $(SUBJ) \rightarrow \emptyset / (BY\ OBJ)$
$(OBJ \rightarrow (SUBJ)$
Morphological *change:* $V \rightarrow V_{[Part]}$

The use of single-headed arrows and the word "change" indicate that the rule is a generative relation-*changing* rule. In fact, the notion of a "redundancy rule" itself is slightly oxymoronic, since a redundant statement of regularity is not in any normal sense rule-like.

In the remainder of this chapter, a resolution of the paradox of partial productivity is suggested, involving two types of learning mechanisms. The first is a certain type of indirect negative evidence; the second mechanism, presumably working in tandem with the first, draws largely on work by Pinker (1989) and Levin (1993) and the related experimental evidence of Gropen et al. (1989).

5.2   INDIRECT NEGATIVE EVIDENCE

I do not attempt to survey the full range of efforts to suggest that some type of indirect negative evidence is available here (see Bowerman 1988 and Pinker 1989 for detailed discussion of the problem and critiques of many possible solutions), but there is one possibility (raised in Pinker 1981, 1984, and then rejected in Pinker 1989) that deserves further study.

Since we have assumed that no two constructions are entirely synonymous both semantically and pragmatically (cf. chapter 3), it should be possible to find contexts in which a given construction is the most preferred. If the preferred form is *not* used, then the child is able to tentatively infer that that form is disallowed. The inference would have to be tentative, since it is unrealistic to expect speakers to systematically use the most felicitous form in all contexts. However, if the situation repeats itself several times, the child's tentative hypothesis may become a fairly strong conviction. In this way, children would have the opportunity to unlearn certain overgeneralizations.

A simple case that may illustrate this is lexical and periphrastic causatives. It is well known that lexical causatives are used for cases of direct causation, whereas periphrastic causatives may be used for indirect causation (cf. 7.4.2). Therefore, for example, after seeing a magician make a bird disappear, the child may expect to hear a lexical causative as in (8), given that the causation is direct:

(8)   *The magician disappeared the bird.

Instead, however, the child may hear a periphrastic causative:

(9)   Look! The magician made the bird disappear.

The child may now tentatively hypothesize that the lexical causative is unavailable. That is, since the causation is direct, the lexical causative would be preferable if it were an option.

As another example, consider a child's strategy in determining whether a given verb can occur in the ditransitive construction. As noted by Erteschik-Shir (1979), and discussed in section 3.4.2 above, the ditransitive and its prepositional paraphrase with *to* differ in the information structure of the clause. In particular, the ditransitive construction requires that the recipient argument be nonfocused (or "non-dominant" in Erteschik-Shir's terminology) and the transferred entity be focused ("dominant"). Prepositional paraphrases prefer the opposite information structure: the recipient tends to be focused, the transferred entity nonfocused. Both of these generalizations are motivated by the fact that focused information tends to come at the end of the nuclear clause.

If the recipient is nonfocused and the transferred entity is focused, we find the ditransitive more acceptable than the prepositional paraphrase:

(10)   a.   Sally gave him a brand-new red Volkswagen. >
       b.   Sally gave a brand-new red Volkswagen to him.

If the recipient argument is focused and the transferred entity nonfocused, we find the reverse situation:

(11)   a.   Sally gave that to a charming young man. >
       b.   Sally gave a charming young man that.

When using verbs which can occur in both constructions, speakers are free to exploit the difference in pragmatic structure. There is, in fact, evidence that children are sensitive to these pragmatic factors (Gropen et al. 1989).

Indirect evidence would arise, then, from situations in which the discourse context matches a certain form but the speaker nevertheless uses a less felicitous form. For example, speakers use the prepositional form for *whisper* even when the information to be conveyed more closely matches the information structure of the ditransitive construction. Thus, if a child hears (12) instead of (13), when the latter might be expected given the fact that the news is the focused information, the child will infer that the ditransitive form is not a possibility for *whisper.*

(12) Sally whispered some terrible news to him.
(13) *Sally whispered him some terrible news.

Pinker raises this possibility in several places (Pinker 1981, 1984:400). However, he ultimately rejects the idea that this mechanism could be sufficient for learning to disallow particular forms, for two reasons (Pinker 1989:16). The first objection he raises is that children's sensitivity to discourse contexts is statistical, not absolute. That is, children do not treat discourse effects as a determinant factor in choosing alternate argument structures; they are more likely to use the argument structure with the better-suited pragmatics, but they do not always do so (Gropen et al. 1989). However, the very fact that children are more likely to use the construction with better-suited pragmatics is sufficient to show that they do have an implicit knowledge of the information structure and are able to attend to it. For instance, it is possible that a child wouldn't notice the first time that *whisper* was used with the focus on the transferred entity, or the second time. But eventually, the child would presumably notice; at that time she would be able to use the input evidence to form the hypothesis that *whisper* cannot be used in the ditransitive form.

Even if we strengthen Pinker's first objection and assume that *adult's* sensitivity to discourse contexts is also only statistical and not absolute, we do not undermine this strategy. That is, even if we concede that neither adults nor children can be assumed to always use the most felicitous form, this strategy is not ruled out. All that is required is that the child be capable of recognizing a statistical correlation in the input data (Kapur 1993).

The second objection Pinker raises stems from his assumption that other focusing devices such as pronouns, cleft constructions, and contrastive stress can be used to override the default differences in information structure between alternative argument structures. On that view, adult speakers would be able to compensate for using less-preferred argument structures by overlaying these less-preferred argument structures with various focusing devices (e.g., pro-nouns, focus constructions, and stress), thus altering the information structure encoded by particular argument structures as a default. Assuming that speakers make use of these strategies, the input would be for the most part optimal, and children would have no reason to infer that the speaker in a given situation would have used a different argument structure if he could have.

However, this suggestion is ultimately not persuasive, since focusing devices are not able to alter or override the information structure of the clause but instead are required to obey the independently existing information structure of the clause. For example, pronouns are preferred in nonfocus positions:

(14)   a.   She gave it to a woman. >
      b. *She gave a woman it.

(15)   a.   She gave her a brand-new house. >
      b. #She gave a brand-new house to her.

Similarly, focus has been argued to only pick out arguments that are in focus-able positions as defined for a given construction. For example, as mentioned above, Erteschik-Shir (1979) has argued that the recipient argument of the di-transitive construction is not available as focus because the construction re-quires that argument to be nonfocused (or "non-dominant"):

(16)   a.??Who did you give the book?
      b.??It was Mary you gave the book.
      c.??Was it Mary you gave the book?

Finally, stress is also more felicitous on arguments that are in focus position, thereby generally emphasizing the information structure rather than overriding it. Thus example (17a) is more felicitous than (17b):

(17)   a.   She gave that to A WOMAN SHE JUST MET. >
      b. #She gave A WOMAN SHE JUST MET that.

Therefore, focusing devices might well be limited to giving the child addi-tional evidence for the information structure of the clause, rather than serving to dilute other evidence by providing ways for the adult speaker to circumvent the information structure associated with a particular argument structure. Since two constructions generally differ either semantically or pragmatically, the hy-pothesis that indirect negative evidence is inferred from hearing a verb in a less-than-optimal construction deserves further study.

5.3   CIRCUMSCRIBING VERB CLASSES

Pinker (1989), arguing against any negative evidence (direct or indirect), ultimately provides a different, compelling resolution of the paradox of partial productivity. A broad-range rule is proposed to capture the necessary conditions for a verb's occurrence in additional syntactic frames. In the case of the ditransitive, Pinker posits a broad-range rule that states in effect that a "prospective possessor" must be involved—that is, the first object referent must be understood to be a prospective possessor. This general rule does not provide sufficient conditions, however, there being many verbs that can be understood to involve a prospective possessor which do not allow ditransitive syntax (e.g., *donate, contribute, pull, shout, choose, credit, say*).

Drawing on work by Green (1974) (and Levin 1985 and Rappaport & Levin (1985) for the locative alternation), Pinker suggests that sufficient conditions are determined by a set of narrow-range rules which classify verbs into narrowly defined semantic classes. The specific classes that Pinker proposes are the following (cf. also Gropen et al. 1989; Levin 1993):

    1. Verbs that inherently signify acts of giving: e.g., *give, pass, hand, sell, trade, lend, serve, feed*

    2. Verbs of instantaneous causation of ballistic motion: e.g., *throw, toss, flip, slap, poke, fling, shoot, blast*

    3. Verbs of sending: e.g., *send, mail, ship*

    4. Verbs of continuous causation of accompanied motion in a deictically specified direction: *bring, take*

    5. Verbs of future having (involving a commitment that a person will have something at some later point): e.g., *offer, promise, bequeath, leave, refer, forward, allocate, guarantee, allot, assign, advance, award, reserve, grant*

    6. Verbs of communicated message: e.g., *tell, show, ask, teach, pose, write, spin, read, quote, cite*

    7. Verbs of instrument of communication: e.g., *radio, e-mail, telegraph, wire, telephone, netmail, fax*

    8. Verbs of creation: e.g., *bake, make, build, cook, sew, knit, toss* (when a salad results), *fix* (when a meal results), *pour* (when a drink results)

    9. Verbs of obtaining: e.g., *get, buy, find, steal, order, win, earn, grab*

It may seem that if we admit the possibility of indirect negative evidence as suggested above, there is no need to adopt Pinker's suggestion that narrowly defined semantic classes also play a role in the acquisition of argument structure. However, circumscribing narrowly defined classes of verbs to be associated with a particular construction will allow us to account for extremely low-frequency or novel non-alternating verbs (since the assumed notion of indirect negative evidence presupposes hearing the verb in a non-optimal construction on several occasions).[2] For example, taking Zwicky's example of a novel verb *greem,* defined as a manner-of-speaking verb referring to speech of a loud, hoarse quality, speakers presumably know that they *cannot* say (18):

(18) *He greemed her the news.

This knowledge cannot be attributed to any kind of indirect negative evidence, because the verb is novel; speakers would not have had a chance to unlearn or avoid learning it in this use. Other situations in which the type of indirect negative evidence suggested above would not be an aid to acquisition might include cases in which the construction in question is so low-frequency that the child can never with any modicum of confidence expect its occurrence, and cases in which there is no construction which is closely enough related semantically to the target construction so that the child would be able to infer that the speaker would have used the target construction if possible.[3] Moreover, the generalizations Pinker describes are real; it is necessary to account for the fact that verbs which are used in particular argument structures do often fall into similarity clusters (cf. Levin 1993). There is no reason not to believe that children exploit multiple sources of evidence for learning argument structure; it is suggested here that they make use of narrowly defined verb classes as well as appealing to some degree of indirect negative evidence as described above.

Before continuing with Pinker's argument, we might make several small comments on the particular set of subclasses he proposes, listed above. The fifth subclass, "verbs of future having," can be seen to conflate three distinct subclasses. Some of the verbs are used in expressions which imply that the subject argument actually *acts* to cause the first object argument to receive the second object argument at some later point in time (e.g., *bequeath, leave, forward, allocate, assign*). In other cases, only if the *conditions of satisfaction* (Searle 1983) associated with the act denoted by the predicate hold does the subject argument cause the first object argument to receive the second object argument at some later point in time (e.g., *promise, guarantee, owe*). Finally, some verbs are used in expressions which imply that the subject argument only *enables* the first object argument to receive the second object argument (e.g., *permit, allow*) (cf. discussion in section 3.3.2).

The sixth class, "verbs of communicated message," should be understood to include verbs whose inherent semantics involves a communicative act, in order to distinguish this class from similar verbs such as *say, assert, claim,* and *doubt* which might be described as verbs of propositional attitude. Understood in this way, several of the verbs listed by Pinker seem to be misclassified; for example, *pose* and *spin* do not obviously fall into the class of "verbs of communicated message," and accordingly (at least in my dialect) are not readily dativizable:

(19)   a. ?*Bill posed him a problem.
        b. ?*Bill spun her a fairy tale.

Both this class and the seventh class, "verbs of instrument of communication," should be classified as metaphorical classes since they are based on a systematic metaphor that involves understanding communicated information as being linguistically packaged and exchanged between interlocutors (Reddy 1979).

Finally, at least one additional subclass should be added to the list, namely, verbs of refusal such as *refuse, deny.* Expressions involving these verbs, like (20a, b), imply that the subject argument refuses to cause the first object argument to receive the second object argument.

(20)   a.   Bill refused Joe a raise.
        b.   The committee denied him a promotion.

In any case, we need only accept the spirit of Pinker's analysis—that there is a need to identify narrowly defined semantic subclasses—in order to accept his conclusion that this type of narrow circumscription allows us to capture the fact that subclasses of verbs which refer to the same kinds of general events as the ones listed, but do not fall into any of the above particular classes, fail to dativize. His examples of such nondativizing classes are as follows:

1. Verbs of fulfilling (X gives something to Y that Y deserves, needs, or is worthy of): *I presented him the award; *I credited him the discovery. *Bill entrusted/trusted him the sacred chalice; *I supplied them a bag of groceries.* [I would also include in this class *concede, furnish,* and *donate.*]

2. Verbs of continuous causation of accompanied motion in some manner: *I pulled/carried/pushed/schlepped/lifted/lowered/hauled John the box.*

3. Verbs of manner of speaking: *John shouted/screamed/murmured/ whispered/yodeled Bill the news.*

4. Verbs of proposition and propositional attitudes: *I said/asserted/ questioned/claimed/doubted her something.*

5. Verbs of choosing: *I chose/picked/selected/favored/indicated her a dress.*

Gropen et al. (1989) provide experimental evidence to show that speakers are sensitive to certain morphophonological constraints. In particular, verbs with particular morphemes such as *per-, con-, -mit, -sume* and polysyllabic verbs with non-initial stress are disallowed from participation in the ditransitive construction. These constraints largely coincide with distinctions between Latinate and native vocabulary, and between specialized and more basic vocabulary; however, we clearly would not want to ascribe recourse to etymological information to children, and the experiments in support of these particular constraints controlled for semantic information. Therefore, the constraints are stated in terms of morphophonology. They are used to explain the following:

(21)   Chris bought/*purchased/*obtained/*collected him some food.

(22)   Jan told/*explained/*reported/*announced Chris a story.

However, the constraints do not apply to every narrowly defined class of verbs. Verbs of future having, in particular, are not subject to them:

(23)   Chris assigned/allotted/guaranteed/bequeathed him the tickets.

The class of instrument-of-communication verbs and the class of creation verbs also include verbs which are exceptions to the morphophonological constraints:

(24)   Chris e-mailed/radioed/arpanetted him a message.

(25)   Chris xeroxed/thermofaxed/nroffed him a copy.

Gropen et al. suggest that each of the verbs in (24–25) is classified, independently of the morphological criteria, as a special kind of complex stem having a noun or name as its root. They cite evidence that tacit knowledge of a word's stem being from another syntactic category allows it to be treated specially with respect to morphological processes (cf. Pinker & Prince 1988). To account for these cases, we can state the generalization that a verb from any class which is understood to have a noun or name as its root is not constrained by the morphophonological constraints.

The narrowly defined subclasses of verbs together with the morphophonological constraints provide a high degree of predictive power. A new or nonsense verb which falls into one of the recognized narrow classes of verbs and which, if applicable, obeys the morphophonological constraints is automatically licensed to be used ditransitively (but see the next section). Verbs in conflict with these requirements are ruled out. This circumscription of narrow domains in which the ditransitive is productive goes a long way toward accounting for the apparent paradox that Pinker set out to resolve: that the ditransitive syntax can be extended to new and novel verbs, but at the same time is not available to all verbs of any broadly defined class.

5.4   EXCEPTIONS

The above generalizations are compelling, and in fact every researcher who has studied the semantics of the ditransitive construction in any detail has found it necessary to classify verbs which occur in the construction as belonging to narrowly defined subclasses as a descriptive device (cf. Green 1974;

Oehrle 1976; Wierzbicka 1986). Still, there are various kinds of exceptions to the generalizations just described. First, there are a couple of members in some subclasses which do occur, yet the subclasses are not fully productive. Second, there is at times a certain degree of variability in judgments for verbs which are supposedly within the same narrowly defined class. Finally, there are exceptional verbs such as *envy* and *forgive* which do occur in the ditransitive construction although they do not entail the relevant semantics.

Each of these cases is discussed in turn. In section 5.3, an interpretation of the nature of the verb classes is suggested which can naturally account for all of these seemingly problematic phenomena.

5.4.1   Unproductive Subclasses

The small classes of verbs of permission (*permit, allow*) and verbs of refusal (*refuse, deny*) are unique in not forming productive subclasses:

(26)   Sally permitted/allowed/*let/*enabled Bob a kiss.
(27)   Sally refused/denied/*prevented/*disallowed/*forbade him a kiss.

These classes actually have a slightly different status in the theory proposed by Pinker (1989), because the verbs in these classes do not alternate with prepositional paraphrases. Thus on Pinker's account, these semantically related verbs are not eligible to undergo the lexical rule. However, since we are not postulating a lexical rule, we cannot appeal to the same solution. We need another way to account for their lack of productivity.

5.4.2   Differences in Judgment within Classes

An expected source of idiosyncrasy stems from the fact that the determination of the narrowly defined class which a given verb belongs in is not always entirely clear-cut. For example, I have suggested that *bequeath* falls into the dativizing class of verbs of future having, along with *leave, forward, allocate,* etc. However, it seems that on semantic grounds it might be equally plausible to instead classify *bequeath* in the nondativizing class of verbs of fulfilling (X gives something to Y that Y deserves, needs, or is worthy of), along with *present, credit, entrust, donate,* etc. Because of these two classification possibilities, we would expect *bequeath* in fact to dativize in some dialects and not to dativize in others. In general, in the case of verbs that on the basis of their meaning may fall into one of two classes, one which can appear ditransitively and one which cannot, we would expect to find some dialectal variation as to whether these verbs can be used ditransitively.

Another source of lexical idiosyncrasy is evidenced by the fact that speakers occasionally report different degrees of grammaticality even among verbs which are uncontroversially within the same narrow-range class. For example, *throw* and *blast* both fall within the class of "verbs of instantaneous causation of ballistic motion" (Pinker 1989; cf. above), yet (28) is decidedly better for many speakers than (29):

(28)   She threw him a cannonball.
(29) *She blasted him a cannonball.

Similarly, (30) is judged to be more grammatical than (31), although both *design* and *create* should fail to dativize because of the verbs' non-initial stress.

(30)   ?Sally designed him a sculpture.
(31)??Sally created him a sculpture.

These facts are not obviously accounted for on Pinker's proposal, since according to his theory the productive rule should operate blindly within narrowly defined classes; there is no reason to expect some instances to be judged more acceptable than others.[4]

5.4.3 Positive Exceptions

There are a few ditransitive expressions that do not entail any associated transfer. Some uses of *ask* can be fit into the pattern described above if they are interpreted as instances of the metaphor involving understanding information as traveling from speaker to hearer (cf. Reddy 1979). For example, (32) can be understood to mean that Amy caused Sam to "receive" a question:

(32)   Amy asked Sam a question.

However, other uses of *ask* are clearly exceptional; consider (33):

(33)   Amy asked Sam his name/his birthday/his marital status.

This type of example clearly does not imply that Sam potentially receives his name, his birthday, or his marital status. Grimshaw (1979) discusses these "concealed questions" at some length. She argues that noun phrases such as those above, which are questions semantically, can appear as arguments of any verb which subcategorizes for an NP in that position and which selects for a question complement. Thus example (33) is motivated by factors that are independent of the ditransitive construction, resulting in a case of "target-structure conspiracy" in the sense of Green (1973).[5]

Forgive and especially *envy,* as used in (34) and (35), respectively, are also exceptional:

(34)   He forgave her her sins.
(35)   He envied the prince his fortune.

The subjects in these cases are not causal, and no reception is involved. However, these predicates have illuminating semantic histories. *Forgive* and *envy* historically had senses that were closely related to *give*. *Forgive* used to mean "to give or grant" (OED). *Envy* used to mean "to give grudgingly" or "to refuse to give a thing to" (OED). This of course is not evidence that *forgive* or *envy* are part of the synchronic semantic pattern outlined above. But the historical facts do suggest that these predicates were at least at one time associated with this sort of pattern. Correspondingly, these facts also suggest that a construction can occasionally be frozen without continuing reference to the original semantics.

However, it seems reasonable that syntactic change should tend toward patterns that are more transparent to the speaker. If the construction with the semantics outlined here is psychologically real, then it would be natural for odd cases of ditransitives involving *forgive* and *envy* to drop out of use. I myself find archaic-sounding sentences involving *forgive* and *envy* much more acceptable than modern-sounding sentences; for example:

(36)   a.   She forgave him his sins.
         b.?*She forgave him his goof.
(37)   a.   She envied him his vast fortune.
         b.?*She envied him his extensive stock portfolio.

And in fact, other speakers are even less accepting of these constructions. In attempting to explain the idea of positive exceptions to a class of undergraduate cognitive science students, I wrote sentence (36a) and (37a) on the board. In response an audible groan arose from the class. When asked what was wrong, the students said they didn't find those sentences acceptable (this judgment was held by more than half of them). Thus it seems indeed that *envy* and *forgive* are dropping out of the language (at least among speakers under twenty-one), just as we would expect if the semantics associated with the ditransitive pattern were synchronically real.

Nonetheless, *envy* and *forgive* have been exceptions for some time, and have been learned as such by generations of speakers. Thus an adequate account of grammar must allow for some degree of lexical idiosyncrasy, despite the powerful effect of semantic motivations (cf. also Lakoff 1965; Fillmore 1977b; Rosen 1984; Mithun 1991; Dowty 1991). Note that these cases are unproblematic from the point of view of learning, since the child has positive evidence that the verbs in question are used in the ditransitive construction, and can therefore learn them on an instance-by-instance basis as idioms.

5.5 ACCOUNTING FOR THE EXCEPTIONS: A USAGE-BASED ACCOUNT
5.5.1 Productivity Defined by Verb Clusters

In the preceding sections, we have seen that even after embracing the idea of narrowly defined verb classes to account for the partial productivity of the ditransitive construction, there remains a residue of lexical idiosyncrasy. There are small subclasses which are not productive, varying degrees of acceptability within seemingly productive subclasses, and positive exceptions to the semantic generalizations, such as *envy* and *forgive*.

This idiosyncrasy is in fact expected if one considers certain experimental findings. Gropen et al. (1989) suggest that speakers "tend to be conservative" in their use of lexical items. Specifically, they show that people tend to use lexical items in the same constructions in which they have heard those items used by others, but that they can, if properly primed, extend the uses to new patterns.[6] This phenomena would be impossible if people did not store in memory the specific syntactic patterns that a word is heard used with (see also Bybee 1985 and Langacker 1987a for particular usage-based models of grammar). This being the case, a certain degree of lexical idiosyncrasy is to be expected.

However, the existence of some degree of lexical idiosyncrasy should not be taken as counterevidence against the existence of narrowly defined semantic subclasses of verbs that occur in the ditransitive construction. Although the exact formulation of these classes has differed, their existence as such has been recognized by every researcher who has looked in any detail at the verbs occurring in this construction. And, as has just been discussed (and is spelled out in more detail in Pinker 1989), the existence of such classes helps to explain the phenomena of partial productivity.

These two facts—that there are narrowly defined productive verb classes, and that at the same time we find scattered positive exceptions and varying degrees of acceptability within these narrowly defined classes—can be reconciled by recognizing verb classes to be *implicitly* represented speaker-internally as generalizations over learned instances. Because memory is associative, similar verbs used in the same constructions are classified together by general categorization processes. Therefore the claim is that speakers attempt to *categorize* learned instances.

Narrowly defined verb classes, then, are implicitly represented as clusters of semantically related verbs known to occur with a given construction. New or previously unclassified verb forms are attracted to existing clusters on the basis of similarity to existing cases. However, judgments of similarity are notoriously variable across speakers and contexts, and two activities can almost always be said to be similar in some respect. Therefore, in order to adequately defend the idea that the use of new and novel senses is determined by similarity to existing cases, one must be able to define the similarity metric which is to be used as the basis of comparison. On the present account, the characterizations of the verb classes themselves can be viewed as providing a similarity metric. For example, if one of the verb classes associated with the ditransitive is "verbs of ballistic motion," then we can consider *shin* to be relevantly like *kick* in that it is a verb of ballistic motion.

The determination of which verb classes are relevant, or alternatively, what features of similarity are important, requires empirical crosslinguistic study, and I do not claim to provide an account here (but cf. Pinker 1989 and Levin 1993 for discussion and suggestions, and Mufwene 1978 for an early discussion on this subject). Only by looking at which distinctions are made crosslinguistically can we determine what the semantically (or morphophonologically) relevant aspects of verb meaning are that determine the basis of the clustering into subclasses.

On this view, frequency is expected to affect the classification of new verbs. Two kinds of frequency information need to be distinguished. On the one hand there is *token frequency,* which refers to the number of times a given instance (e.g., a particular word) is used in a particular construction; on the other hand there is *type frequency,* which refers to the number of distinct words that occur in a particular construction. MacWhinney (1978) and Bybee (1985) have argued that it is the type frequency of a particular process (or a particular construction) that plays a crucial role in determining how likely it is that the process may be extended to new forms: the higher the type frequency, the higher the productivity.

To see the relevance of the type/token frequency distinction for productivity, consider the following example cited by Bybee (1985:132–133). She notes that Guillaume (1927) documented the fact that French-speaking children most frequently overgeneralize the use of first-conjugation suffixes with verbs of other conjugations. He also observed the number of verbs of each conjugation used spontaneously in children's speech. Bybee cites the following table, which shows the number of occurrences of each conjugation class and the number of verbs used from each class:

Conjugation Class	Number of Uses	Number of Verbs
First (*chanter*)	1,060 (36.2%)	124 (76.0%)
Second (*finir*)	173 (6%)	10 (6.1%)
Third (*vendre*)	1,706 (57.8%)	29 (17.9%)

Although more than half of the number of tokens ("uses") of verbs were of the third conjugation, the number of different verbs that occurred in this class was much smaller than the number that occurred in the first conjugation. Correspondingly, the first conjugation was seen to be much more productively used.

The proposal to implicitly represent verb classes as similarity clusters can perhaps be made more clear by the following rough-and-ready representation (which does not take morphophonological similarity into account):

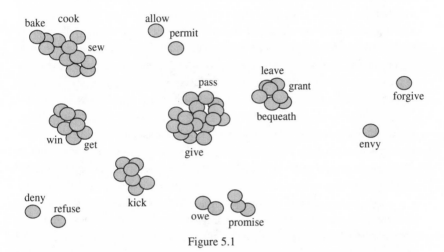

Figure 5.1

Each circle represents a lexical entry; the entries are projected onto two dimensions, with semantically closer verbs being represented by physically closer circles. By way of demonstration, one or more instances within a given cluster have been labeled. The circles representing *bake* and *cook*, for example, are close together to indicate their being in the same narrowly defined class.[7]

Type frequency can be discerned by considering the number of circles in a given cluster. Clusters containing more circles are more likely to be productive. Subclasses with only two members such as the verbs of refusal (*deny, refuse*) are expected not to be productive, because of their low type frequency.

The idea that verbs are represented this way in an associative memory is of course inspired by recent connectionist representations. However, the diagram need not be construed as necessarily presupposing a connectionist model of memory; all that is needed is an associative memory (e.g., as proposed in work in the domain of morphology of Pinker and Prince (1991)).

This view of the way new verbs are attracted to learned instances makes several predictions. First, it predicts that subclasses with few members will not contain enough instances to create a similarity class, and so will not be productive. Secondly, it predicts the possibility of differences in judgments within similarity classes. Such differences will result from (1) the degree of similarity between the case being judged and other cases within the subclass, and (2) the relative type frequency that the relevant cluster displays. It is not necessary (or possible) to exhaustively list all the verbs that can potentially occur in a given construction. Novel cases are analogized to previously learned cases on the basis of their similarity to these familiar cases and the type frequency of these familiar cases.

Occasional positive exceptions (such as *envy* and *forgive* for the ditransitive construction) are tolerated because speakers simply associate the words with the constructions idiosyncratically. There is no danger of productive extensions from these outliers because they, like subclasses of fewer than two members, do not constitute a cluster, and therefore do not attract novel cases.

The representation in figure 5.1 entails that the knowledge that certain verbs are used in a particular construction is part of a speaker's competence. However, it is not necessary that each new entry be stored as an additional member of a cluster, throughout the speaker's life. It is possible that once a critical number of instances in a particular cluster is learned—insuring that novel instances that fall into the class will be included—new cases are no longer stored in memory since they would provide only entirely redundant information. It is also possible that learned instances are not necessarily stored as discrete, clearly individuated cases; rather, the edges of learned instances that form a cluster may blend into each other, delimiting an area in semantic space without specifically retaining each individual instance within.

5.5.2   Varying Degrees of Productivity

The ditransitive provides a good example of a construction with associated verb classes. The degree of productivity of other constructions can be seen to form a cline between those constructions which are not fully productive even within narrowly defined verb classes and those which approach full productivity as long as general constraints are obeyed. An example of the first case, that of very limited productivity (at least in some dialects), is the resultative construction. As discussed in chapter 8, there is a large degree of idiosyncrasy as to which verbs can occur with which resultatives. Notice the following contrasts:

(38) a. She shot him dead. >
      b.??She blasted him dead.

(39) a.   She cried herself to sleep. >
      b.   She cried herself asleep. >
      c.??She wept herself to sleep/asleep.

(40) a.??He ate himself asleep.
      b.??He cried himself sick.

At the same time, it is clear that resultatives are not entirely idiomatic and do occasionally occur productively. For example:

(41) a.   "I cried myself well-nigh blind." (Tennyson, "Grandmother X" (1884); cited by Visser 1963)
      b.   "Drive your engine clean" (Mobil ad; cited by Rappaport Hovav & Levin 1991)
      c.   "She could wonder herself crazy over the human eyebrow." (R. L. Stevenson, "Virginibus Puerisque" (1881))

The particular factors which underlie the limited productivity of this construction must include semantic factors of the type outlined in chapter 8. In addition, morphophonological factors, such as the ones Gropen et al. (1989) found to be relevant in the case of the ditransitive, and the token frequency of the analogical source may need to be taken into account. The role of each of these factors remains an issue for further research.

An example at the opposite end of the continuum is the *way* construction discussed in chapter 9. This construction appears to be almost entirely productive. The following examples come from the Oxford University Press corpus:

(42) a.   "But he consummately ad-libbed his way through a largely secret press meeting."
      b.   ". . . nasty gossip about me now sludging its way through the intestines of the society I know . . ."
      c.   ". . . their customers snorted and injected their way to oblivion and sometimes died on the stairs."
      d.   ". . . [they] hoped they too could massage their way to keeping power."
      e.   "Lord King craftily joked and blustered his way out of trouble at the meeting."

As discussed in chapter 9, the few non-occurring verbs (such as the vanilla motion verbs *go, walk, move*) can be accounted for by general semantic constraints on the construction. Interestingly, the token frequency of this construction is low, with one example occurring approximately every 40,000–56,000 words in the Lund Corpus of conversational texts and the Wall Street Journal.[8] This provides support for the idea that productivity has little to do with token frequency and more with type frequency.

The range of differing productivity illustrated by these examples is exactly what we would expect given the usage-based model of grammar described above. That is, learned and thus stored resultative cases are few and only dot the semantic landscape; little or no clustering of examples is attested. Therefore novel extensions *sound* novel and are not fully idiomatic, unlike such extensions of the ditransitive cases as *fax* or Marantz's novel verb *shin.*

At the same time, attested *way* construction examples seem to span the spectrum of semantic space, given the general constraints imposed by the construction. Since the construction has such a high type frequency of attested verbs, novel verbs are freely used in it.

5.5.3   Productive Links

As discussed in chapter 3, relations between constructions, represented by various types of inheritance links, are also objects in our system. Different instances of a given link occur with different type frequencies, just as different instances of a particular construction occur with different type frequencies.

For example, the causative–inchoative relation, which is represented by a kind of subsumption inheritance link, occurs between the caused-motion and intransitive motion constructions, the resultative and the intransitive resultative constructions, and the simple causative and simple inchoative constructions. Therefore this particular link would be said to have a type frequency of at least three.

Some of the polysemous extensions we have seen occur in both the ditransitive construction and the caused-motion construction. The type frequency of each of these polysemy links is increased with every construction which is extended in the same way. Because productivity is directly correlated with type frequency, the higher the type frequency, the more likely a particular inheritance link will exist between pairs of new constructions that are relevantly similar to the pairs of existing constructions which the inheritance link already relates. In the limiting case, a link will apply fully productively, yielding extensions every time a novel construction is encountered, as long as that construction satisfies the particular semantic characteristics of the existing instances. In this case, the link between the two constructions is quite analogous to a rule, in that the existence of one form can be used to predict the existence of the other form.

For example, the passive construction, discussed briefly at the end of chapter 2, is instantiated by many different particular versions, each corresponding to an active construction with the relevant semantics (the active construction must have at least two arguments, with one being higher on the role hierarchy than the other). Because the type of link between active and passive constructions occurs between so many different active and passive pairs, it has an extremely high type frequency. Therefore the passive link is, in effect, rule-like in its application.

5.6 CONCLUSION

The account proposed here to explain the partial productivity of constructions involves two types of learning mechanisms. The first is a type of indirect negative evidence, based on the hypothesis that every construction contrasts with every other construction. Therefore, upon hearing a verb in a construction that must be considered non-optimal given the current context, the learner tentatively hypothesizes that the verb cannot occur in what would be the preferred construction. The reasoning is roughly, "If that construction could have been used, I guess it would have been used; therefore maybe it can't be used." Upon witnessing the verb in a non-optimal construction, given the context, a number of times, the learner's hypothesis that the verb cannot occur in the optimal construction is strengthened. This strategy was first proposed by Pinker (1981); his later rejection of this strategy (Pinker 1989) was argued here to have been unwarranted.

The second learning mechanism, presumably working in tandem with the first, draws largely on recent work by Pinker (1989) and the related experimental evidence of Gropen et al. (1989). Specifically, the need to circumscribe narrowly defined semantic subclasses characterized by local productivity is acknowledged.

The account proposed here differs somewhat from Pinker's and Gropen's in that on the present account, the narrowly defined subclasses are understood to be clusters defined by semantic and morphophonological similarity that are conventionally associated with the construction, as opposed to subclasses that are conventionally allowed to undergo a lexical rule. Moreover, on the account presented here, the verb classes are interpreted as implicit generalizations over learned instances in order to account for small nonproductive subclasses, differences in judgments even within narrowly defined classes, and the existence of positive exceptions such as *envy* and *forgive*. Viewing verb classes as clusters of cases in an associative memory also allows us to assimilate other constructions which involve markedly more or less productivity. In particular, it was shown that we might actually expect the fact that the resultative construction is productive only to a limited degree, whereas the *way* construction is almost fully productive.

It may seem that by allowing the knowledge of whether a verb is used in a particular construction to be stored, we undermine the existence of the construction as an independent entity. That is, if we need to posit the fact that *kick* can be used with the ditransitive construction as a separate piece of grammatical knowledge, why not instead posit a new sense of *kick,* along the lines suggested by semantics-changing lexical rule accounts (cf. the discussion in chapter 4)?

The reason for postulating constructions is analogous to the reason why other researchers have wanted to postulate a lexical rule: in order to capture generalizations across instances. Moreover, it is claimed here that what is stored is the knowledge that a particular verb *with its inherent meaning* can be used in a particular construction. This is equivalent to saying that the composite fused structure involving both verb and construction is stored in memory. By recognizing the stored entity to be a *composite* structure, we gain the benefits described in chapter 1 over a lexical rule account. For example, we avoid implausible verb senses such as "to cause to receive by kicking." It is the *composite* structure of verb and construction that has this meaning. We also allow other syntactic processes to refer to the inherent lexical semantics of the verb. Thus we do not lose the information conveyed by the verb, because the verb is not changed into a new verb with a different sense.

6 The English Ditransitive Construction

6.1 Introduction

The ditransitive construction has already been discussed with respect to its polysemy (chapter 2) and its partial productivity (chapter 5). In this chapter, I concentrate on particular semantic constraints and metaphorical extensions of the construction. Highly specific semantic constraints are associated directly with the ditransitive argument structure, revealing a more specific semantic structure than is generally acknowledged. In particular, the central sense is argued to involve transfer between a volitional agent and a willing recipient. Several systematic metaphors are identified and associated with the construction, showing that expressions such as *Mary gave Joe a kiss* and *Mary's behavior gave John an idea,* which are often assumed to be idiosyncratic, are instances of a large and productive class of expressions that are based on systematic metaphors.

Before getting to those constraints, however, evidence that a construction is indeed required in this case is reviewed.

6.2 The Existence of the Construction

Following the program laid out in previous chapters, we need to show that aspects of the syntax or semantics of ditransitive expressions are not predictable from other constructions existing in the grammar. First, to see that the construction contributes semantics not attributable to the lexical items involved, consider the verb *bake* when used ditransitively:

(1)   Sally baked her sister a cake.

This expression can only mean that Sally baked a cake with the intention of giving it to her sister. It cannot mean that Sally baked the cake so that her sister wouldn't have to bake it; nor can it mean that Sally baked the cake as a demonstration of cake-baking, or that she baked a cake for herself because her sister wanted her to have one. Unless we associate the "intended transfer" aspect of meaning to the construction, we are forced to say that *bake* itself means something like 'X intends to cause Y to receive Z by baking.' This "transfer sense" of *bake* would be posited only to avoid attributing aspects of the semantics to the construction. The positing of such ad hoc verb senses which only occur in a particular construction was argued against extensively in previous chapters.

In addition, as was noticed by Partee (1965:60) and Green (1974:103), the goal argument of ditransitives must be animate—that is, it must be a recipient:

(2)   a.   She brought the boarder/*the border a package. (cited by Gropen et al. (1989), attributed to J. Bresnan)

As has been argued in chapters 1 and 4, this semantic constraint is most parsimoniously attributed to the construction.

Ditransitive expressions are syntactically unique in allowing two nonpredicative noun phrases to occur directly after the verb; the fact that English will allow such a configuration is not predictable from other constructions in the language. In addition, this is the only construction which links the recipient role with the OBJ grammatical function.

The construction was represented in figure 2.4, repeated here as figure 6.1.

Ditransitive Construction

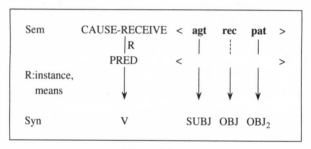

Figure 6.1

The construction's agent and patient roles must be fused with independently existing participant roles of the verb (represented by the PRED variable), as is indicated by the solid lines between the agent and patient argument roles and the predicate's participant role array, which is unfilled in the above diagram. The recipient role may be contributed by the construction; this is indicated by the dashed line between the recipient argument role and the array of predicate participant roles.

6.3   THE SEMANTICS

The semantics of the ditransitive construction has not been understudied, and this work owes a large debt to previous analyses, in particular to Cattell (1984), Green (1974), and Oehrle (1976) for their detailed discussion of hundreds of ditransitive expressions.

6.3.1   Volitionality of the Agent

There are certain semantic constraints on the ditransitive syntax which have not been incorporated into most theories of argument structure. The reason these constraints are often overlooked is that there appear to be exceptional cases. However, the exceptional cases form a delimitable class that can be seen to involve a general systematic metaphor (of the type described in Lakoff & Johnson 1980). It will be shown that the constraints do in fact hold in the source domain of the metaphor.

To identify the first constraint, notice that each of the verbs described so far independently selects for a volitional subject argument. This generalization can be captured by assigning a constraint on the nature of the subject argument directly to the construction.

The volitionality must extend so that not only is the action described by the verb performed agentively, but also with the relevant transfer intended. For example, in (3) below, Joe must be understood to intend to give the picture to Sally. It cannot be the case that Joe painted the picture for someone else and later happened to give it to Sally.

(3)   Joe painted Sally a picture.

Similarly, in (4) it cannot be the case that Bob told the story to someone else and Joe just happened to overhear.[1]

(4)   Bob told Joe a story.

This constraint also accounts for the ill-formedness of the following examples:

(5)   *Joe threw the right fielder the ball he had intended the first baseman to catch.
(6)   *Hal brought his mother a cake since he didn't eat it on the way home.
(7)   *Joe took Sam a package by leaving it in his trunk where Sam later found it.

This is not to say that the first or second object arguments of the ditransitive cannot be given a transparent interpretation. The description used to pick out the argument referents may be understood to be the speaker's description, not the subject argument's. For example, consider (8):

(8)   Oedipus gave his mother a kiss.

This sentence is felicitous despite the fact that Oedipus did not realize he was kissing his mother. Likewise for (9):

(9)   Joe gave Mary a sweater with a hole in it.

This statement is acceptable even if Joe did not intend to give Mary a defective sweater. Also, it is not necessarily contradictory to use "accidentally" in ditransitive expressions; for example:

(10) Joe accidentally loaned Bob a lot of money [by mistaking Bob for Bill, his twin; without realizing that Bob would skip bail with it; instead of giving the money as a gift as he had intended].

While I do not attempt to untangle the relevant issues here, I appeal to the fact that the same possibilities of interpretation occur with other expressions that are generally agreed to require volitional subject arguments. For example, *murder* is a verb which is universally recognized as selecting for a volitional subject argument. Still, it is possible to say the following without contradiction:

(11) Mary accidentally murdered Jane [although she had meant to murder Sue; although she had only meant to knock her unconscious].

What I am suggesting, then, is that whatever notion of volitionality is adopted to deal with verbs such as *murder* should also be used to capture the semantic requirement of the subject position of the ditransitive construction.

The existence of this constraint has been obscured by examples such as these:

(12) a. The medicine brought him relief.
      b. The rain bought us some time.
      c. She got me a ticket by distracting me while I was driving.
      d. She gave me the flu.
      e. The music lent the party a festive air.
      f. The missed ball handed him the victory on a silver platter.

In these examples the subject argument is not volitional. Even when the subject argument is an animate being, as in (12c, d), no volitionality is required. However, these examples form a delimitable class of expressions, as they are all instances of a particular conventional systematic metaphor, namely, "causal events as transfers."[2] This metaphor involves understanding causing an effect in an entity as transferring the effect, construed as an object, to that entity. Evidence for the existence of this metaphor independent of the ditransitive construction comes from the following expressions:

(13) a. The Catch-22 situation *presented* him with a dilemma.
      b. The unforeseen circumstances *laid* a new opportunity *at our feet*.
      c. The document *supplied* us with some entertainment.
      d. The report *furnished* them with the information they needed.

Further evidence, both for the existence of the metaphor and for it motivating the ditransitive examples in (12), comes from the polysemy of each of the predicates involved in those examples. The predicates *bring, buy, get, give, lend,* and *hand* are used to imply causation, but on their basic sense they each involve transfer from an agent to a recipient. The link between these senses is provided by the metaphor. *Bring, buy, get, give, lend,* and *hand* here involve the metaphorical transfer of effect; each of the examples in (12) implies that the subject argument is the cause of the first object argument being affected in some way by "receiving" the second object argument.

This class can be represented as an extension of the central sense as follows:

Ditransitive Construction

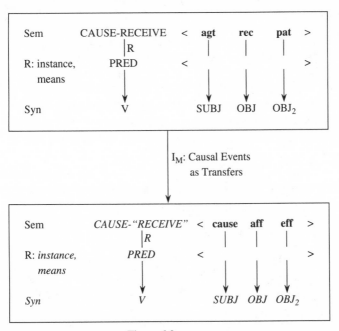

Figure 6.2

Recognizing the metaphor allows us to divorce ourselves from the often-made but erroneous claim that examples such as those in (12) are idiosyncratic.

Returning to the statement of the constraint that the subject argument must intend the transfer, we can see the necessity of acknowledging the role of this metaphor. It is this metaphor which licenses the exceptional cases: we can recognize that the volitionality constraint is satisfied in the source domain of the metaphor. At the same time, this metaphor differs from other metaphors to be described below in not mapping volitionality to the target domain. This follows from the fact that the target domain is concerned with abstract causes. Abstract causes cannot be necessarily volitional because they are not necessarily human. Each of the metaphors described below, on the other hand, involves human actors in the target domain as well as in the source domain, and in each of the target domains the volitionality constraint is respected.

6.3.2   Semantic Constraints on the Recipient

As noted above, it has long been realized that the referent designated by the first object must be an animate being. However, this constraint, just like the constraint that the subject argument must intend the transfer, is somewhat obscured by expressions licensed by the causal-events-as-transfers metaphor. Consider (14–16):

(14) The paint job gave the car a higher sale price.

(15) The tabasco sauce gave the baked beans some flavor.

(16) The music lent the party a festive air.

In none of these examples is the first object an animate being; however, in the source domain of the metaphor the affected party is understood to be a recipient, and thus indeed an animate being. Again we find that a constraint can be satisfied in the source but not the target domain of the metaphor.

An additional semantic constraint is that the first object be understood to be a beneficiary, or a *willing* recipient.[3] This constraint is needed to account for the following example from Green (1974):

(17) *Sally burned Joe some rice.

Example (17) is unacceptable even if malicious intentions are attributed to Sally; however, it *is* acceptable in the context that Joe is thought to like burnt rice. Furthermore, one cannot felicitously say either of the following:

(18) *Bill told Mary a story, but she wasn't listening.

(19) *Bill threw the coma victim a blanket.

In these examples, the first object is not understood to be a willing recipient; accordingly, they are unacceptable.

This constraint may also be responsible for the slight difference in meaning between the following two examples provided by Robert Wilensky (personal communication):

(20) a.   She fed lasagna to the guests.
      b.   She fed the guests lasagna.

Most speakers find the first example to be somewhat less polite than the second. Since *feed* is normally used with reference to the food intake of babies or animals, the impoliteness of the first example is not surprising; what requires explanation is the fact that the second example is interpreted to be relatively more polite. The constraint that the first object must be construed as a willing recipient can account for this, since the ditransitive version has the effect of imposing the interpretation that the guests are willing agents, thereby according them more respect.

That the recipient is expected to be willing should not be confused with the idea that the recipient is expected to benefit from the transfer. Thus, while (21) below does not imply that Jane will benefit from imbibing the martini, it does presuppose that she is expected to willingly drink the martini.

(21) Jack poured Jane an arsenic-laced martini.

In some cases, however, the issue of the recipient's willingness or unwillingness is irrelevant to whether transfer is successful. These involve expressions in which actual successful transfer is implied:[4]

(22) Bill gave the driver a speeding ticket.
(23) Bill gave Chris a headache.
(24) Bill gave Chris a kick.

Nonetheless, all cases in which the first object is required to *accept* the transferred object in order for transfer to be successful imply that the first object is assumed to be a willing recipient.

6.3.3   On the Notion "Recipient"

I have been referring to the semantic role of the first object position as "recipient" instead of as "goal" or "possessor." In view of the above constraint—that this argument be animate—"recipient" is clearly more accurate than "goal." At the same time, "recipient" is preferable to "possessor" because many of the metaphors involving transfer (to be described below) do not map the implication that the recipient actually possesses the transferred entity after reception. Consider (25):

(25) Jo gave Mary an insult.

This sentence does not imply that Mary "possesses" an insult, only that Mary "received" an insult. Similarly with (26):

(26) Jan gave Chris a punch.

This example does not imply that Chris "possesses" a punch but only that he "received" one. If we describe the role in question as that of "recipient" instead of "possessor," these facts pose no problem. The fact that a possessive relationship is usually implied follows automatically from the fact that what is received is normally subsequently possessed.

Noticing that a recipient is involved in ditransitive expressions may be a first step toward motivating the double object syntax of the construction. Beginning with Jakobson, those interested in the semantics of the direct object have noted that *recipients* of force and effect make for good direct objects (Jakobson 1938; for recent discussion see, e.g., Langacker 1987;[6] Rice 1987a). (Of course this is not to say that all direct objects are recipients; clearly the objects of cognition verbs such as *believe, see,* and *know* would present difficulties for such a claim.)

Finally, the construction has been shown to be associated with a scene of transfer. Describing the first object as a "recipient" rather than "possessor" more adequately captures the dynamic character of this semantics.

6.3.4   Other Metaphors

The systematic metaphor of causal events as transfers is just one of several metaphors which license the use of the ditransitive construction. Other metaphors can be understood to license other extensions. The source domain of each of these metaphors is the central sense of actual successful transfer.

The "conduit metaphor," described and named by Michael Reddy (1979), involves communication *traveling across* from the stimulus to the listener. The listener understands the communication upon "reception." Evidence for the metaphor includes:

(27) a. He *got the ideas across to* Jo.
      b. His thoughts *came across from* his speech.
      c. Jo *received* the information from Sam.
      d. Jo *got the information from* Bill.

This metaphor licenses the following examples:

(28) She told Jo a fairy tale.

(29) She wired Jo a message.

(30) She quoted Jo a passage.

(31) She gave Jo her thoughts on the subject.

A related metaphor involves understanding perceptions as entities which move toward the perceiver. The perception is understood to occur upon "reception." Evidence for the metaphor includes the following:

(32) a. The view *knocked me over.*
      b. I *caught* a glimpse of him.
      c. I *missed* that sight.
      d. I *had* a view of the orchestra.
      e. He *let me have* a look.

This metaphor licenses the following examples:

(33) He showed Bob the view.

(34) He gave Bob a glimpse.

Another metaphor involves understanding actions that are intentionally directed at another person as being entities which are transferred to that person. Evidence for the metaphor includes:

(35) a. He *blocked* the kick.
      b. He *caught* the kiss she *threw to* him.
      c. All he *got from her* was a goodbye wave.
      d. Joe *took a punch from* Bill.
      e. She couldn't *get a smile out of* him.
      f. She *threw* a parting glance *in his direction.*
      g. She *targeted him* with a big smile.
      h. Bob *received* a slap/kick/kiss/smile from Jo.

This metaphor licenses the following expressions:

(36) She blew him a kiss.

(37) She shot him a keep-quiet look.

(38) She gave him a wink.

(39) She gave him a punch.

(40) She threw him a parting glance.

Another metaphor extends the use of the ditransitive to the speech act domain. This metaphor is used in reference to a situation where a person insists on certain facts and assumptions. The metaphor involves understanding these facts and assumptions as objects which are given to someone who is making an argument, to be used in the construction of the argument. (The idea of *constructing an argument* assumes yet another metaphor, that of arguments as constructed objects). We can title this metaphor "Facts and Assumptions as

Objects which are Given." Evidence for the metaphor includes the following expressions:

(41) a. I'll let you *have* that much.
      b. I don't want to *give up* that assumption.
      c. *Accept* that as a *given.*
      d. If you *take* that assumption *away,* you don't *have* a great argument.
      e. If you don't *have* that assumption, you're not *left* with much.
      f. Even *granted* that, your argument is still full of holes.

This metaphor licenses the following:

(42) I'll give you that assumption.

(43) I'll grant you that much of your argument.

The final metaphor to be discussed here licenses ditransitive expressions which are often assumed not to involve a possessor at all. The following examples come from Green 1974:

(44) Crush me a mountain.

(45) Cry me a river.

(46) Slay me a dragon.

(47) They're going to kill Reagan a commie.

These expressions can be seen to involve metaphorical transfer once the following metaphor is recognized: actions which are performed for the benefit of a person are understood as objects which are transferred to that person. The metaphor is exemplified in the following expressions:

(48) a. He *owes* you many favors.
      b. By slaving away quietly for him, she has *given* more than he deserves.
      c. The senator claimed never to have *received* any favors.
      d. He always *gets* what he wants out of people.
      e. She graciously *offered* a ride to the airport.

The mapping of this metaphor is different from that of the others in that the source domain of this metaphor is not 'X CAUSES Y to RECEIVE Z' as it was for each of the others. In particular, it is the action performed rather than the second object argument that is the received object in the mapping. This metaphor, then, represents an extended use of the ditransitive. And, as we might expect, there is wide dialectal variation in the degree of acceptability of these expressions. In fact, these cases are subject to their own special constraints. As noted by Oehrle (1976), they are more acceptable as commands:

(49) a. Cry me a river.
      b. ?Sally cried me a river.

They are also more acceptable with pronouns in first object position. Contrast (49a) with (50):

(50) ?Cry Joe a river.

To summarize, these cases can be seen to be a limited extension from the central sense of the construction. The source domain of this metaphor is not 'X CAUSES Y to RECEIVE Z' as it was in each of the other metaphors; rather, it is 'X CAUSES Y to RECEIVE an OBJECT (not necessarily designated Z).' The target domain is 'X Performs an Action for the Benefit of Y.' Z is mapped to the object acted on by X.

6.4 CONCLUSION

In this chapter the central sense of the ditransitive construction has been argued to be associated with a highly specific semantic structure, that of successful transfer between a volitional agent and a willing recipient. In addition, several systematic metaphors that license extensions from the basic sense have been identified.

7 The English Caused-Motion Construction

7.1 INTRODUCTION

In this chapter the "caused-motion" construction is discussed in some detail, and arguments that a particular construction of this kind is required in the grammar are made explicit. In particular, it is argued that such a construction must be specified in the grammar to account for certain cases in which the semantic interpretation cannot plausibly be attributed to the main verb and other means of deriving the semantics compositionally also fail.

This construction can be defined (in active form) structurally as follows (where V is a nonstative verb and OBL is a directional phrase):

[SUBJ [V OBJ OBL]]

This definition is meant to cover the following types of expressions:

(1)   They laughed the poor guy out of the room.

(2)   Frank sneezed the tissue off the table.

(3)   Mary urged Bill into the house.

(4)   Sue let the water out of the bathtub.

(5)   Sam helped him into the car.

(6)   They sprayed the paint onto the wall.

The basic semantics of this construction as described in chapter 3 is that the causer argument directly causes the theme argument to move along a path designated by the directional phrase; that is, 'X CAUSES Y to MOVE Z'. The various extensions from this basic sense will be discussed in more detail below.

Quite specific semantic constraints on the types of situations that can be expressed by this construction are proposed, revealing principled patterns where there is apparent idiosyncrasy. Each of these constraints intuitively falls under the heading of "direct" causation, or under what can count as a single event; therefore the constraints can be viewed as beginning to provide a more specific characterization of these notional terms.

A discussion of the classic *load/spray* alternation follows in section 7.6. It is argued that the majority of verbs showing this alternation—by occurring in both of the constructions below—can be accounted for without positing additional verb senses.

(7)   a.   Pat loaded hay onto the truck.
       b.   Pat loaded the truck with hay.

7.2   THE EXISTENCE OF THE CONSTRUCTION

Again, in order to show that a distinct construction is required, it is necessary to show that its semantics is not compositionally derived from other
constructions existing in the grammar. For example, it is necessary to show
that some aspect of the construction is not compositionally derived from the
lexical items (i.e., lexical constructions) which instantiate it. It is also necessary
to show that the particular combination of lexical items does not inevitably lead
to the particular interpretation in question. In this section, attempts to derive
the semantics of the caused-motion construction from the verb's inherent semantics, the preposition's semantics, or the semantics of the combination of
verb plus preposition are critically discussed.

Several observations in the literature lead to the conclusion that the verb in
isolation does not inherently encode the caused-motion semantics. As Fillmore
(1971), Talmy (1976), and Randall (1983) have noticed, many verbs are not
causative verbs independently of this construction. For example, *kick* and *hit*
in (8) and (9) do not have causative interpretations:

(8)   Joe kicked the wall.
(9)   Joe hit the table.

Yet, when these verbs are used in the caused-motion construction, a causal
interpretation is implied:

(10)   Joe kicked the dog into the bathroom.
        ($\rightarrow$ He caused the dog to move into the bathroom)
(11)   Joe hit the ball across the field.
        ($\rightarrow$ He caused the ball to move across the field)

As has been noticed by Aske (1989), it is also the case that many verbs do not
necessarily code motion independently of this construction. Aske provides the
following contrast, and notes that it is implausible to posit a distinct motion
sense for the predicate *squeeze:*

(12)   a.   Frank squeezed the ball.
              ($\nrightarrow$ The ball necessarily moves.)
        b.   Frank squeezed the ball through the crack.
              ($\rightarrow$ The ball necessarily moves.)

Also, as has been noted by Green (1973), Randall (1983), and Hoekstra (1988),
many transitive verbs which can occur in this construction do not bear the same

semantic relation to their direct object as they do in simple transitive sentences. Consider (13a–c):

(13) a. Sam sawed/tore/hacked/ripped a piece off the block.
      b. Sam rinsed/cleaned the soap out of her eyes.
      c. Sam stirred the paint thinner into the paint.

These do not entail (14a–c):

(14) a. Sam sawed/tore/hacked/ripped a piece.
      b. Sam washed/rinsed/cleaned the soap.
      c. Sam mixed/stirred the paint thinner.

Verbs can sometimes appear in this construction that do not independently license direct object complements at all:

(15) The audience laughed the poor guy off of the stage.

(16) Frank sneezed the napkin off the table.

(17) In the last Star Trek episode, there was a woman who could think people into a different galaxy.

The move to postulate novel causative motion senses for each of these verbs, thereby positing the meaning of the whole in the meaning of the parts by stipulation, has been argued against in chapter 1. Several other proposals to account for the caused-motion interpretation by compositionally deriving the meaning from the combination of verb and preposition are critically discussed below.

◆

Several authors have proposed accounts that avoid positing rampant lexical polysemy to account for the caused-motion construction and the related resultative construction (see chapter 8). Gawron (1985, 1986), Pustejovsky (1991a), Rappaport Hovav and Levin (1991), and Hoekstra (1992), for example, have argued that the meanings of resultatives and/or caused-motion expressions such as those in examples (1–6) above do not require positing additional verb senses. These authors argue that the meaning of the entire sentence can be compositionally derived from composing the meanings of the constituent parts.

A general problem with compositional accounts such as these stems from the far-reaching conclusion drawn from the fact that we may be able to pragmatically infer the meaning of the construction. If one knows that a construction has a particular form, then it is sometimes the case that one may reasonably infer that it has the particular interpretation that it has. However, it is fallacious to argue that because we may be able to pragmatically infer the meaning of a construction, its existence is therefore predictable rather than conventionalized. Such reasoning is based solely on a model of interpretation, yet we also must account for production. For instance, while we may be able to infer that expressions of the form exemplified in (1–6) have the semantics they do, we cannot predict that constructions of this type will exist. In fact, Talmy (1985a) points out that this pattern of expression is not available as a productive form in Romance, Semitic, or Polynesian language families, although it does occur in Chinese as well as in Dutch and English.[1]

Makkai's (1972) distinction between "idioms of encoding" and "idioms of decoding" can be used to make the point. Decoding idioms are idioms which a listener would be unable to confidently interpret without having learned the idiom separately. Encoding idioms are idioms whose meaning may be inferable; all the same, without having heard the idiom a speaker would have no way of knowing that it was a conventional way of saying what it says. *Fly by night, by and large* are examples of decoding idioms; *serial killer, sofa bed* are examples of encoding idioms. By referring to both kinds of terms as *idioms,* Makkai makes the point that neither kind is predictable from general pragmatic principles. Still, it is worth examining the compositional accounts mentioned above on their own terms.

Gawron (1985, 1986) argues that caused-motion expressions consist of two predicates—a verb and a preposition—and that both of these retain their normal meanings. The relation between the two predicates, if not determined independently by the verb's semantics, is said to be pragmatically inferable from the possible relations that can hold in general between predicates in a single clause. A new way for complements to be semantically conjoined with verbs is introduced to account for some expressions similar to those in examples (1–7): "*co-predication.*" The verb and the preposition act as co-predicators, sharing one argument and combining semantically in pragmatically inferable ways. For example, *John broke the hammer against the vase* is analyzed roughly as 'Break(John,the-hammer), Against(the-hammer, the-vase)'. The preposition *against* is claimed to be responsible for the interpretation that the hammer comes into forceful contact with the vase.[2]

Pustejovksy's (1991a) account is relevantly similar. He suggests that the verbs involved in caused-motion or resultative expressions are lexically transitive process verbs that are combined with independent PPs (or, in the case of resultatives, APs). The PP is associated with its own event structure: that of a state. The composition process-plus-state is claimed to inevitably yield a transition (accomplishment) interpretation, namely that of a caused motion or caused change of state.

Pustejovsky claims that the PP or AP is an adjunct. However, the syntactic

attachment of the PP or adjective is not that of an adjunct. This is true regardless of whether the PP is sister to the verb as claimed here or is part of a small
clause that is sister to the verb. Moreover, it is not possible to attribute adjunct
status to the result phrase and claim that the causative interpretation is inferred,
because there exist cases which do not receive a causative interpretation. These
are the well-known depictive predicates as well as standard PP adjuncts.[3]

(18) Depictives
   a. The witch-hunters burned her *alive.*
   b. Sam passed Bob the towel *wet.*
(19) Adjunct PPs
   a. Lisa slept *under the bridge.*
   b. Joe played *in the house.*

A problem for both Gawron's and Pustejovsky's approach is the existence
of caused-motion expressions that involve predicates which cannot occur transitively at all. For example, we cannot account for (16) repeated here as
(20), in terms of 'Sneeze(Fred,the-napkin), Off(the-napkin,the-table)', because
the first predication—'Sneeze(Fred,the-napkin)'—is nonsensical (cf. also examples (15, 17)).

(20) Fred sneezed the napkin off the table.

This type of example is not discussed by Pustejovsky or Gawron. Presumably,
both accounts would require a three-argument sense for verbs such as *sneeze,*
thereby resorting to the kind of polysemy they had otherwise sought to avoid.

Hoekstra's (1992) account actually focuses on the type of resultative found
in examples (13), (20), and those below:

(21) a. Fred mixed the paint thinner into the paint.
   b. Ethel washed the soap out of her eyes.

Notice that in examples (21a, b) the verb does not bear its normal relation to
the direct object complement. Hoekstra analyzes such resultatives as involving
an intransitive process verb combined with a small clause stative predicate.[4]
This is similar to a proposal made by Rappaport Hovav and Levin (1991) to
account for this type of resultative. They also propose that the verb is lexically
intransitive—in particular, that it is unergative—and that it is combined with
an independent small clause complement.[5]

Both Hoekstra's and Rappaport Hovav and Levin's analyses require some
way to join the main verb and the small clause. They cannot simply say that
the verb selects for (or theta-marks) the small clause complement, since this

would imply an additional sense for each verb, something they wish to avoid. For example, they wish to avoid positing a new sense of *drink* that has an agent and a state argument to account for (22):

(22)   She drank him under the table.

Rappaport Hovav and Levin suggest that the verb case-marks the NP of the small clause. But case marking normally applies to NPs inside small clause *arguments* of the verb. Since Rappaport Hovav and Levin do not wish to claim that the small clause is an argument, they do not, as far as I can tell, explain what licenses the occurrence of the small clause itself. To account for the resultative interpretation, Rappaport Hovav and Levin suggest that because the small clause XP is attached at the lowest bar level within the VP, it has to be semantically integrated into the "core eventuality" named by the verb. However, normally only lexically-subcategorized-for complements are attached at the lowest bar level within the VP; therefore the semantic generalization hinges on a syntactic constraint that is left unexplained.

Hoekstra proposes a new way to syntactically compose the verb and small clause. To do this, he needs to rely on rather nonstandard and often seemingly unmotivated assumptions. A critical assumption of his is that points of time associated with verbs are theta-marked. In the case of an activity predicate like *drink,* a final time point t_n may be theta-marked through binding the "e-position" of the small clause which denotes a state.[6] The resultative interpretation is claimed to be determined by the way in which the small clause is licensed, namely, through the final point in the event structure of the matrix verb (p. 162). However, Hoekstra claims that the resultative can only occur with atelic activity verbs;[7] a priori it would seem that there is no "final point" of time associated with atelic predicates. Hoekstra does not make clear how or why it is that such a final point is available.

To summarize, there is a general problem with attempts to account for caused-motion and resultative expressions by means of a simple concatenation of two independently existing predicates, the semantic interpretation being arrived at by general pragmatic principles. Such analyses do not account for the fact that such concatenation is allowed in the language in the first place. Unless we treat the secondary predicate as an argument or an adjunct, there are no preexisting means by which to concatenate the two predicates. Treating the secondary predicate as an argument of the matrix verb is tantamount to creating an extended sense of the verb, a move that all of these accounts wish to avoid and which has been argued against in chapter 1. Treating the secondary predicate as an adjunct is not a viable solution either, because its syntactic position (whether sister to the verb or part of a small clause) is not that of an adjunct; moreover, the semantic interpretation is different than that of other adjunct secondary predicates such as depictive predicates and PP adjuncts. Hoekstra's suggestion of a new way to join verb and secondary predicate is not well motivated, and Gawron's proposal requires additional verb senses for some cases and non-argument PP phrases for others. In the latter case, the burden of the change in meaning falls onto the preposition. This raises the question whether the preposition in itself could be held responsible for the caused-motion semantics (cf. also Aske 1989).

A major problem with attributing the burden of the semantic interpretation of caused motion to the preposition is that many prepositions which appear in this construction favor a locative interpretation:

(23) Fred stuffed the papers *in* the envelope.

(24) Sam pushed him *within* arm's length of the grenade.

(25) Sam shoved him *outside* the room.

It might be suggested that these prepositions are systematically ambiguous in English, being able to receive either a locative or a directional interpretation. However, such a proposal fails to account for the intuition that prepositions such as *inside, in, outside,* and *within* do *not* intuitively code motion on either use. More to the point, these terms are not ambiguous in all contexts. For example, when fronted they can only receive a locative interpretation:

(26) a. Into the room he ran, quick as lightning.
      b. *Inside the room he ran, quick as lightning. (on the directional reading that he ran into the room)
      c. *Within the room he ran, quick as lightning. (on the directional reading)

An account relying on an ambiguity of these terms would need to specify in exactly which contexts the ambiguity could arise.

In response to the possible suggestion that we might attribute the motion interpretation to *either* the verb *or* the preposition, but that one or the other must lexically specify motion, consider examples such as the following:

(27) Sam squeezed the rubber ball inside the jar.

(28) Sam urged Bill outside of the house.

In these cases, neither the verbs *squeeze* or *urge* nor the prepositions *inside* or *outside* independently code motion.

Therefore, since the causal interpretation cannot be systematically attributed to either the verb or the preposition or their combination, and since systematically attributing the motion interpretation to the preposition requires that seemingly locative prepositions such as *within* are said to be ambiguous although they are demonstrably not ambiguous in other contexts, we attribute the caused-motion interpretation to a construction which combines the verb and directional preposition yielding a particular, conventionalized interpretation.

The possibility of allowing "basically" locative, nondirectional PP's in this construction raises a question for our account, since we have specified that the construction must contain a PP coding a *directional* phrase. However, we noted that terms which are intuitively locative cannot receive a directional interpretation in all contexts. What needs to be recognized to account for these cases is a particular process of *accommodation* (cf. Talmy 1977; Carter 1988) or *coercion* (Moens & Steedman 1988; Croft 1991; Pustejovsky 1991b; Sag & Pollard 1991) by which the construction is able to *coerce* the locative term into a directional reading.

On the view taken here, coercion is not a purely pragmatic process; rather, it is only licensed by *particular constructions* in the language. That is, coercion is only possible when a construction requires a particular interpretation that is not independently coded by particular lexical items. To the extent that the occurring lexical items can be coerced by the construction into having a different but related interpretation, the entire expression will be judged grammatical.[8] On this view, the locative terms are not independently ambiguous, but instead are capable of being coerced *by particular constructions* into having the related directional meaning. In the case at hand, we can understand the locative terms to be coerced into having a directional meaning by the caused-motion construction itself.

In order for coercion to be possible, there needs to be a relationship between the inherent meaning of the lexical items and the coerced interpretation. Clearly it is not possible for just any lexical item to be coerced into receiving a directional interpretation. The relationship between the meaning of the locative term and the directional interpretation it receives is one of *endpoint focus* (Brugman 1988). That is, the location encoded by the locative phrase is interpreted to be the endpoint of a path to that location.

To summarize, because attempts to attribute the meanings of entire expressions of caused motion to the meanings of individual lexical items fall short of accounting for the data in a natural way, a construction is posited in the grammar. The construction can be represented as follows:

Caused-Motion Construction

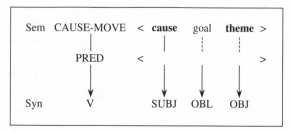

Figure 7.1

A distinct but related construction must be posited to account for intransitive motion cases; this construction can also add a motion interpretation to verbs that do not lexically code motion:[9]

(29)   The bottle floated into the cave. (Talmy 1985a)

The intransitive motion construction is related to the caused-motion construction as follows:

Caused-Motion Construction

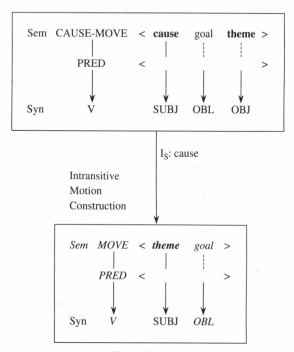

Figure 7.2

7.3   THE VARIOUS INTERPRETATIONS

7.3.1 Overview

As discussed in chapter 3, the caused-motion construction is associated with a category of related senses. The various senses that occur are the following:

A. **'X CAUSES Y to MOVE Z'.** Examples include:

(30)   Frank pushed it into the box.

(31)   Frank kicked the dog into the bathroom.

(32)   Frank sneezed the tissue off the nightstand.

(33)   Sam shoved it into the carton.

B. **The conditions of satisfaction associated with the act denoted by the predicate entail: 'X causes Y to move Z'.** Included in this class are force-dynamic verbs (Talmy 1985b) that encode a communicative act. Examples of this class are the following:

(34)   Sam ordered him out of the house.

(35)   Sam asked him into the room.

(36)   Sam invited him out to her cabin.

(37)   Sam beckoned him into the room.

(38)   Sam urged him into the room.

(39)   Sam sent him to the market.

These examples differ from the previous cases in that motion is not strictly entailed. For example, Sam ordering someone out of the house does not necessarily entail that the person moves out of the house. However, motion *is* entailed by the "conditions of satisfaction" (Searle 1983) associated with the actions denoted by the particular predicates. If in example (34) the order is *satisfied,* the person will leave the house. Similarly, if the request in (35) or invitation in (36) is satisfied, the person will move along the designated path.

C. **'X ENABLES Y to MOVE Z'.** This class includes force-dynamic verbs that encode the removal of a barrier, such as *allow, let, free, release.* For example:

(40)   Sam allowed Bob out of the room.

(41)   Sam let Bill into the room.

In general, "enablement" is understood force-dynamically to involve either the active removal of a barrier or the failure to impose a potential barrier (Talmy 1976; Sweetser 1990). However, this construction allows only the former type of enablement, whereby the enabler retains some aspect of agentivity. That is, enablement that does not actively involve the removal of a barrier is not acceptable in caused-motion expressions:

(42) a. *Sara let Bill into the room by leaving the door open.
        b.   (Sara let Bill come into the room by leaving the door open.)

D. **'X PREVENTS Y from MOVING Comp(Z)'.** This class of expressions, by contrast to the one above, can be described in terms of the force-dynamic schema of imposition of a barrier, causing the patient to stay in a location despite its inherent tendency to move. The class includes verbs such as *lock, keep, barricade*. For example:

(43) Harry locked Joe into the bathroom.

(44) He kept her at arm's length.

(45) Sam barricaded him out of the room.

The path argument of this class, argument Comp(Z), codes the *complement* of the potential motion. Thus (43) implies that Harry prevented Joe from moving *out of* the bathroom.

E. **'X HELPS Y to MOVE Z'.** This class of cases involves ongoing assistance to move in a certain direction. Examples include:

(46) Sam helped him into the car.

(47) Sam assisted her out of the room.

(48) Sam guided him through the terrain.

(49) Sam showed him into the livingroom.

(57) Sam walked him to the car.

7.3.2   Relating the Senses

Sense *A* (e.g., *He pushed the box into the room*) can be argued to be the central sense of the construction. It involves manipulative causation and actual movement, the scene to which transitive markers are applied earliest cross-linguistically (Slobin 1985) and which has been suggested as the most basic causative situation (Talmy 1976). Moreover, the other extensions are most economically described as extensions of this sense.

The entire category of related senses can be diagrammed as follows:

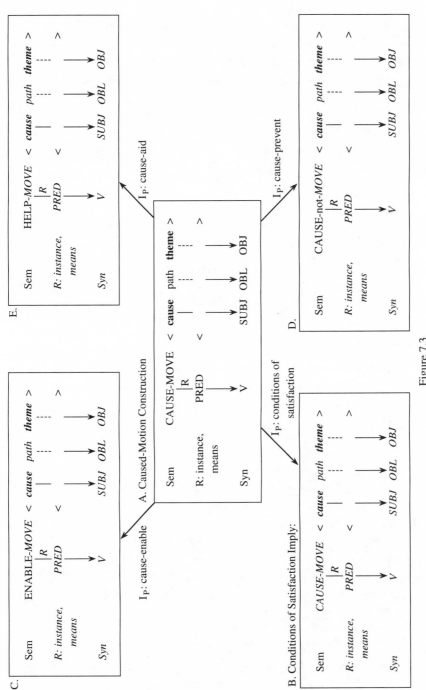

Figure 7.3

The close parallel between the links required for this construction and the links previously suggested for the ditransitive construction indicates that these patterns of extension are quite systematic. It seems that patterns of constructional extension, like patterns of polysemy generally, embody subregularities (Wilensky 1991); that is, patterns of polysemy recur, although not strictly predictably so. At the very least it should be clear that the links are not being posited on an ad hoc basis.

A Further Extension. Finally, there are a few verbs which do not fit into any of the above patterns, in that the subject is not interpreted to cause, enable, or prevent the theme's motion. For example, consider the verb *accompany* as used in (51):

(51)   Sam accompanied Bob into the room.

Here *accompany*, although similar to uses of *escort, walk,* and *show,* does not necessarily entail any assistance on the part of the agent. Also, *follow, trail,* and *tail,* while similar to the unexceptional use of *chase* in (52), differ in that they do not entail that the theme's motion is caused or aided by the agent.

(52)   Ann chased the squirrel out of her house.

This entire group of exceptions can be characterized as involving a subclass of verbs which entail that the agent argument as well as the theme argument move along the specified path. Thus this class can be recognized as a further extension of the pattern.

7.4   SEMANTIC CONSTRAINTS

In section 7.2 we argued for the existence of the caused-motion construction independently of the verbs which instantiate it. One of our primary motivations for doing so was to avoid arbitrary lexical stipulations on each verb that could potentially occur in the construction. Still, if we were to find that there is rampant lexical idiosyncrasy associated with the construction, our motivation for postulating it would be partially undermined, since the arbitrary lexical stipulation we were trying to avoid would then be necessary anyway. Therefore it is worthwhile to see how much can be accounted for in a principled way by paying close attention to semantic constraints.

At first glance, there does appear to be a large degree of idiosyncrasy. Consider the following minimal pairs:

(53)   a.   Pat coaxed him into the room.
       b. *Pat encouraged him into the room.

(54) a. He hit the ball over the fence.
  b. *He struck the ball over the fence. (adapted from Jackendoff 1990a)
(55) a. Please chop the kindling into the bin provided for it.
  b.??Please don't chop the kindling onto the rug. (Paul Kay, personal communication)
(56) a. Pat asked him into the room.
  b. *Pat begged him into the room.

However, in what follows it will be argued that each of these pairs can be accounted for in a principled way, once careful attention is paid to the semantics of the construction.

7.4.1   A Constraint on the Causer Argument

There is a particular constraint on the causer argument of the caused-motion construction. The cause argument can be an agent or a natural force:

(57) *Chris* pushed the piano up the stairs.

(58) *The wind* blew the ship off course.

(59) *The rain* swept the ring into the gutter.

But it cannot be an instrument (cf. Gawron 1986 for a similar constraint):

(60) a. *The hammer broke the vase into pieces.
  b. *The hammer broke the vase onto the floor.
  c. (The hammer broke the vase.)
(61) a. *His cane helped him into the car.
  b. (His cane helped him get around.)

The fact that the choice of argument encoded as subject plays a role in the acceptability of caused-motion expressions demonstrates that the semantics of the construction must make reference to that argument, and that it cannot be stated as a VP construction.

7.4.2   Constraints on Direct Causation

It has long been recognized that there is a difference in interpretation between lexical causatives such as *kill* and *melt* and periphrastic causatives such as *cause to die* and *cause to melt* (Fodor 1970; Shibatani 1973, 1976; Lakoff 1977; McCawley 1978; Gergely & Bever 1986). Lexical causatives have been argued to primarily involve causation that is "prototypical" (Lakoff 1977), "efficient" (Gawron 1985, 1986), or "direct" (Fodor 1970; Shibatani 1973; McCawley 1978). It has also been suggested that single-clause causative expressions can only express a single event, via an iconic principle (e.g.,

Haiman 1980). In this chapter, specific constraints on what kinds of situations can be encoded by the caused-motion construction are discussed with an aim at explicating these notions. Each of these constraints intuitively falls under the more general heading of "direct causation," or alternatively under the principle of "one event per clause," but each provides a more specific principle by which to characterize such notions.

No Mediating Cognitive Decision

The first thing we might try to explain is the difference in judgments between the following two examples:

(62) a.   Sam coaxed Bob into the room.
      b. *Sam encouraged Bob into the room.

Notice that *convince, persuade, instruct* pattern like *encourage* in not appearing in the caused-motion construction:

(63) *Sam convinced/persuaded/encouraged/instructed him into the room.

Each of these verbs *can* occur in a different construction with an infinitival complement:

(64) Sam convinced/persuaded/encouraged/instructed him to go into the room.

What all of these verbs have in common is that they entail that the entity denoted by the direct object makes a cognitive decision. This is in distinction to verbs such as the ones used in (65–67):

(65) Sam *frightened* Bob out of the house.
(66) Sam *coaxed* him into the room.
(67) Sam *lured* him into the room.

Frighten, coax, and *lure,* although referring to psychological states, do not entail the existence of a cognitive decision. That is, they can apply equally well to rodents without any anthropomorphizing:

(68)   Sam frightened/coaxed/lured the mouse out of its hiding place.
(69)?#Sam encouraged/convinced/persuaded the mouse to move out of its hiding place.

One might argue that this is a coincidence—that verbs which occur with this construction just happen not to entail any cognitive decision on the part of the theme argument and that the "constraint" is an epiphenomenon of particular idiosyncratic lexical facts. However, a piece of evidence weighing against such an account and suggesting instead that it is indeed the construction that prohib-its a mediating cognitive decision comes from an examination of verbs which occur in more than one construction, together with the distribution of the ad-verb *willingly.* Lakoff (1970a) noticed that *willingly* can be applied to both logical (i.e., underlying) and surface subjects:

(70)   a.   Chris$_i$ sold the slave willingly$_i$.
       b.   Chris sold the slave$_i$ willingly$_i$.
(71)   a.   The slave$_i$ was sold willingly$_i$ by Chris.
       b.   The slave was sold willingly$_i$ by Chris$_i$.

According to many theories, the direct object argument is the logical subject of the predicative PP. Yet we find that *willingly* cannot apply to the direct object argument of caused-motion expressions:

(72)   a.   *He asked her$_i$ into the room willingly$_i$.
       b.   He asked her$_i$ to go into the room willingly$_i$.

This is generally true of the passive forms of the caused-motion construction as well:[10]

(73)   a.   *She$_i$ was asked into the room willingly$_i$.
       b.   She$_i$ was asked to go into the room willingly$_i$.

A general constraint against a mediating cognitive decision in the caused-motion construction allows us to prevent *willingly* from applying to the theme argument on semantic grounds, despite the fact that the theme argument may well be considered a logical subject.

The constraint can be stated as follows:

> *Generalization I:* No cognitive decision can mediate between the causing event and the entailed motion.

The Implication of Actual Motion

Another source of apparent idiosyncrasy is the following. There are two subclasses associated with the construction that do not strictly entail actual motion. In the first class, only the conditions of satisfaction associated with the act denoted by the verb entail that the theme argument actually moves. Expres-sions in this class include:

(74)   Sam asked him into the room.
(75)   Sam invited him onto the deck.
(76)   Sam urged him into the room.

The second class involves a specific case of enablement: an agent actively removes a barrier to motion. Expressions in this class include:

(77) Sam allowed him into the room.
(78) Sam permitted him into the house.

These classes do seem to allow a cognitive decision on the part of the theme to be implied if the theme argument actually moves, but they can be distinguished from the cases discussed in the previous section in that actual motion is not entailed by the expression.

Notice that alongside the acceptable (79) we find the unacceptable (80) and (81):

(79) Sam asked Joe into the room.
(80) *Sam begged Joe into the room.
(81) *Sam pleaded Joe into the room.

What needs to be noticed in this case is that the theme's ultimate direction must be presumed to be the one determined by the subject; no contrary tendency can be implied. To see this, compare (82a, b):

(82) a. #Sam asked Harry into the jail cell.
      b.   Sam asked Harry to go into the jail cell.

If the theme's motion is not strictly entailed, it must be presumed as a *ceteris paribus* implication that the theme argument will actually move on the path specified. In the case of *beg* or *plead*—or in (82a), in which there are pragmatically given reasons why Harry may not want to go into the jail cell—motion cannot be presumed.

Notice that it is not necessary for the theme argument to actually *want* to move along the specified path, only that it be presumed to do so:

(83) Sam ordered Bob into the jail cell.

We can summarize these observations as follows:

> *Generalization II:* If motion is not strictly entailed, it must be presumed as a *ceteris paribus* implication.

Conventionalized Scenarios

Certain cases seem to flout the general constraint that there can be no intermediate causation. As Shibatani (1973) noticed, activities which are conventionally accomplished in a particular way may be expressed as simple causatives, even when the causation is indirect insofar as there is in actuality an intermediate cause. For example, consider (84–87):

(84)   The invalid owner ran his favorite horse (in the race).

(85)   Chris cut her hair at the salon on University Avenue.

(86)   She painted her house. (when in fact the painters did the painting)

(87)   Farmer Joe grew those grape vines.

It is a conventional way to have one's hair cut to go to a salon, a conventional way to have one's house painted to have professional painters do it, and so forth. That is, simple causatives can be used to imply *conventionalized* causation that may in actuality involve an intermediate cause. It seems that conventionalized scenarios can be cognitively "packaged" in such a way that their internal structure is ignored.

Notice that many of the same scenes described above cannot occur with directionals:

(88) ?? Farmer Joe grew those vines onto his roof.

(89)   *The invalid owner ran his favorite horse onto the field.

However, we would not want to say that these are ruled out because conventional causation is not acceptable in the caused-motion construction, but rather that these scenes as wholes are not conventional. That is, planting and watering is not a conventional way to grow plants *onto the roof,* and arranging for your horse to run in a race is not a conventional way to have your horse run *onto the field.* Expressions which do express indirect but conventional caused motion are allowable in the caused-motion construction as well:

(90)   The company flew her to Chicago for an interview.

This is acceptable since paying for and arranging a ticket for someone else are conventional ways to have someone travel for interviews.

To sum up this constraint:

> *Generalization III:* Conventionalized scenarios can be cognitively packaged as a single event even if an intervening cause exists.

In the next sections, it will be argued that the action denoted by the verb, performed by the causer on the causee, must be understood to completely determine both the effect of motion and the particular path of motion.

The Effect of Motion

The examples below involve a seemingly idiosyncratic difference between the verbs *hit* and *strike* that has been noticed by Jackendoff (1990a):

(91) a.   He hit the ball across the field.
      b. *He struck the ball across the field.

When a wider class of verbs is considered, a pattern emerges. Notice that the verbs *slap, smack, whack, knock* pattern like *hit*, whereas the verbs *assault, sock, spank, clobber, slash, bludgeon, impact* pattern like *strike*. What distinguishes these two classes of verbs of forceful impact is whether the impacted entity is necessarily affected in a way that does not involve motion (cf. Fillmore 1970). All of the verbs of the *strike*-class require that the impacted entity be affected:[11]

(92) a. *With an open hand, the toddler struck the tree.
      b.   The toddler struck his playmate.
(93) a. *The disgruntled player socked the wall.
      b.   The disgruntled player socked the coach.
      c. *Joe assaulted/bludgeoned/impacted the steel block.

The verbs of the *hit*-class, by contrast, which do allow the directional to be specified, allow their direct objects to be either unaffected generally (*hit, slap*) or unaffected except for the particular effect of motion (*knock*).

(94) Sam hit the table.
(95) Sam slapped/smacked the table.
(96) Sam knocked off the lid.

To make this point more clear, consider also the verb *shoot,* which allows either the impacted entity or the trajectory as direct object:

(97) a.   Pat shot Sam.
      b.   Pat shot the bullet.

Notice that when a path argument is present, the direct object can only be interpreted as trajectory; it cannot be viewed simultaneously as trajectory and impacted entity:

(98) *Pat shot Sam across the room. (unacceptable on the interpretation that Pat shot Sam and the bullet forced him across the room)

This example is explained on our account, because if the bullet is understood to penetrate Sam, then Sam is necessarily affected in a way that does not involve motion, and so a path of motion cannot be specified.

The constraint, then, can be tentatively stated as follows:

> If the action denoted by the verb implies an effect other than motion,
> then a path of motion cannot be specified.

This generalization will need to be slightly revised, however, in view of the examples in the following section.

Change-of-State Verbs

Consider the following examples:

(99) The butcher sliced the salami onto the wax paper.

(100) Joey clumped his potatoes into the middle of his plate.

(101) Joey grated the cheese onto a serving plate.

(102) Sam shredded the papers into the garbage pail.

Each of these examples implies a definite effect on the theme argument quite apart from the motion that is implied. However, these change-of-state verbs can be distinguished from the *strike*-class of verbs just discussed, which also entail a definite effect on the direct object argument, in the following way. The action denoted by each of the verbs in (99–102) as performed by the agent argument on the theme argument typically implies some predictable incidental motion. For example, in slicing salami, the salami normally falls away from the slicer; in clumping potatoes into a pile, the potatoes are moved; the act of grating cheese normally implies that the cheese falls away from the instrument used. It is the path of this incidental motion that can be specified by the directional.

It might be observed that it is not a necessary part of the meaning of *slice* that the sliced object necessarily moves. One can imagine a mechanical bread-slicer that slices bread while the bread is contained in a supporting container, not allowing the bread to fall away after being sliced. Similarly, one can imagine a paper shredder which shreds paper that is fixed in place. However, it is clear that in the neutral context, in which the action is done in the conventional way, the action does entail incidental motion. Thus, in order to account for these cases, we have to appeal not to necessary truth conditions holding of the action denoted by the verb, but rather to the conventional scenario associated with the particular act denoted by the verb.

This class of cases is further constrained. Consider the following contrasts:

(103) a. *Sam unintentionally broke the eggs onto the floor.
     b. Sam carefully broke the eggs into the bowl.

(104) a.??Please don't chop that kindling onto the rug. (said to someone who is chopping kindling on a chopping board in the living room)
     b. Please chop that kindling into the bin provided for it. (example from Paul Kay, personal communication)

In the (b)-cases, the ensuing motion is intended and the examples are acceptable. In the (a)-cases, however, the motion is interpreted as unintentional and they are unacceptable.

The relevant generalization seems to be as follows:

> *Generalization IV:* If the activity causing the change of state (or effect), when performed in the conventional way, effects some incidental motion and, moreover, is performed with the *intention* of causing the motion, then the path of motion may be specified.

In understanding this generalization, it is important to realize that the change of state (or effect) must cause incidental motion as a *result,* not that incidental motion is involved as a *means* of causing the change of state. This is important, because the following are unacceptable:

(105)* She filled water into the tub.
(106)* He covered the blanket over Mary.

In these cases, motion is implied in the scenes associated with *fill* and *cover,* insofar as water must move into the tub and the blanket must move over Mary. However, the motion in these cases occurs as the *means* of accomplishing the change of state; it is not an incidental *effect* of the change of state.

The Path of Motion

A constraint related to the previous one can be recognized by considering the following:

(107) a. *He nudged the ball down the incline. (unless there are repetitive nudges)
     b. He nudged the golf ball into the hole.

Example (107a) is unacceptable despite the fact that the nudging of a ball at the top of an incline can cause the ball to roll down the incline.

The crucial fact is that the causal force initiated by the agent argument in this case does not in itself determine the path of motion; gravity is necessary as an intermediary cause. If the causal force initiated by the agent argument does determine the path of motion, the sentence is acceptable:

(108) He shoved the cart down the incline.

Under many circumstances, a specific path is not determined by the activity described; the direction of force only implies that the theme argument moves out of, or away from, its present location. Accordingly, more specific paths cannot be predicated. This observation can account for the following:

(109) a. #They laughed the poor guy into his car.
     b. They laughed the poor guy off the stage.
     c. They laughed the poor guy out of the auditorium.

Similarly, it can explain the following contrast:

(110) a. #Sam frightened Bob under the bed.
     b. Sam frightened Bob out of the house.
     c. Sam frightened Bob away from the door.

The following generalization is claimed to hold:

> *Generalization V:* the path of motion must be completely determined
> by the causal force.

Therefore, while traditionally there has only been a two-way distinction made between "onset" and "continuous" causation (Talmy 1976),[12] what we see here suggests that "onset" may cover two distinguishable types of causation. The first type is that in which the causing event determines the entire path of motion, even though actual physical contact is not maintained over the entire path. This is the only type of onset causation which is acceptable in caused-motion expressions. The second type of onset causation is that in which the causing event initiates motion but does not itself determine the full subsequent path. This type of onset causation is evident in the following:

(111) Joe's nudging the ball at the top of the incline caused the ball to roll all the way down to the bottom.

Which paths count as being "completely determined" is in part a matter of pragmatics. For example, imagine a group of teenagers crowded around a man who is standing by the door of his car waiting for a friend. The teenagers are intimidating the man by making jokes about him and laughing. In this context, it is felicitous to say the following:

(112) They laughed the poor guy into his car.

Similarly, imagine that Sam is playing a game with a child who is lying on the floor next to the bed. The game involves putting on a scary mask and taking it off again. Each time Sam puts on the mask the child predictably shrieks and rolls under the bed in mock fear. In this context, one *can* felicitously say the following:

(113) Sam, stop frightening Bobby under the bed!

In general, if the action is interpreted to be the driving force determining the particular path of motion, the motion can be said to be "completely determined" by the action.

To summarize the constraints that have been argued for in this section:

I. No cognitive decision can mediate between the causing event and the entailed motion.

II. If the caused motion is not strictly entailed, it must be presumed as a *ceteris paribus* implication.

III. Conventionalized scenarios can be cognitively packaged as a single event even if an intervening cause exists (Shibatani 1973).

IV. If the verb is a change-of-state verb (or a verb of effect), such that the activity causing the change of state (or effect), when performed in a conventional way, effects some incidental motion and, moreover, is performed with the *intention* of causing the motion, the path of motion may be specified.

V. The path of motion must be completely determined by the action denoted by the verb.

7.4.3   The Nature of the Constraints on Direct Causation

On the analysis presented here, the difference in directness between "lexical" causatives and periphrastic causatives cannot be attributed simply to the lexical items themselves, since we have argued that many of the lexical items are not causative independently of this construction. For example, it is not possible to attribute direct causation to *kick* if we accept that *kick* does not itself encode any kind of cause. Moreover, many of the verbs that occur in this construction also occur in constructions which do not entail direct causation (or at least do not obey the constraints outlined here). For example, *force, push, ask, invite* can occur with an infinitival complement, and expressions involving infinitives do not necessarily involve direct causation. Thus expressions with infinitival complements *do* allow the theme to make a cognitive decision:

(114) Sam convinced/persuaded/encouraged/instructed him to go into the room.

Also, they do *not* presuppose that the theme will actually move along the specified path:

(115) Sam asked/begged him to go into the jail cell.

Therefore, the constraint of direct causation must be attributed to the caused-motion construction, or more generally, to a principle stating that only direct causation can be expressed within a single clause.

Several of the constraints described here require access to contextual information and general world knowledge combined with certain specifications of particular lexical items. For example, we saw that change of state (or effect) verbs could occur in the caused-motion construction as long as the activity performed was associated with a conventional scenario that implied incidental motion.

Another example of the influence of pragmatic considerations was seen with respect to the constraint that the causal force must completely determine the path of motion. As was noted above, context plays a role in what kinds of actions can "completely determine" a given path.

The fact that a combination of real-world and situational knowledge together with knowledge of lexical specifications plays a role in the possibilities of argument structure has serious repercussions for theories that make a strict division between semantics and pragmatics. The expression of argument structure is generally taken to exclusively involve semantics (if not exclusively syntax), not pragmatics. Yet these cases suggest that pragmatics, in the sense of general world knowledge, does play a crucial role in argument structure (see also Dinsmore 1979; Jackendoff 1983; Langacker 1987a; Zaenen 1991).

7.5   THE *LOAD/SPRAY* ALTERNATION

A number of verbs, including the well-known examples *load* and *spray,* can occur in two alternate syntactic patterns:

(116) a.   Pat sprayed paint onto the statue.
     b.   Pat sprayed the statue with paint.

(cf., e.g., Partee 1965; Fillmore 1968; Anderson 1971; Rappaport & Levin 1988; Gropen et al. 1991). Example (116a) is actually an instance of the caused-motion construction, and in fact can occur with a wide variety of path phrases besides *onto:*

(117) Pat sprayed the paint toward the window/over the fence/through the woods.

As Anderson (1971) first observed, the *with*-variant is associated with some kind of "holistic" effect. That is, (116b) is interpreted to mean that Pat completely covered the statue with paint, or perhaps that Pat vandalized the statue. This observation follows from an analysis which treats (116b) as an instance of the causative construction. On this analysis, the *with*-phrase is an adjunct, closely related to the *with*-phrase of instrumentals (see Rappaport & Levin 1985 for a discussion of the differences between this *with*-adjunct and the instrumental phrase). A causative analysis is also suggested by Gropen et al. (1991), who argue that in examples such as (116b) a change of state is the "highlighted feature of the event" (p. 6).[13]

Pinker (1989:126–127), drawing on work by Rappaport and Levin (1985) suggests five narrowly defined classes of verbs which can occur in both constructions (see chapter 5 for discussion of the relevance of narrowly defined classes):

1. *Slather*-class: simultaneous forceful contact and motion of a mass against a surface: *slather, smear, brush, dab, daub, plaster, rub, smear, smudge, spread, streak* . . .

2. *Heap*-class: vertical arrangement on a horizontal surface: *heap, pile, stack* . . .

3. *Spray*-class: force is imparted to a mass, causing ballistic motion in a specified spatial distribution along a trajectory: *spray, spatter, splash, splatter, inject, sprinkle, squirt* . . .

4. *Cram*-class: mass is forced into a container against the limits of its capacity: *cram, pack, crowd, jam, stuff* . . .[14]

5. *Load*-class: a mass of a size, shape, or type defined by the intended use of a container (and not purely by its geometry) is put into the container, enabling it to accomplish its function: *load, pack* (of suitcases), *stock* (of shelves) . . .[15]

Verbs of the *slather*-class require all three participant roles to be expressed. Notice, for example, that one can say either (118a) or (118b):

(118) a.   Sam slathered shaving cream onto his face.
       b.   Sam slathered his face with shaving cream.

None of the following examples, with one role unexpressed, are allowed:

(119) a.   *Sam slathered shaving cream.
       b.   *Sam slathered his face.
       c.   *Shaving cream slathered onto his face.

This is represented by profiling all three roles:

slather ⟨**slatherer, thick-mass, target**⟩

Slather is compatible with both the caused-motion construction and the causative-plus-*with*-adjunct constructions in the following way. Both constructions allow all three roles to be expressed, so there is no problem satisfying the constraint that profiled roles are obligatory. Since there are three profiled participants, one may be fused with a nonprofiled argument role, in accord with the Principle of Correspondence discussed in section 2.4.2.

The verb's participant roles are fused with each of the construction's argument roles in accord with the Principle of Semantic Coherence, also discussed in section 2.4.2, as follows. The verb's participant roles fuse with the caused-motion construction's argument roles in that the *slatherer* can be construed as a cause, *thick-mass* as a type of theme since it undergoes a change of location, and the *target* as a type of goal-path. *Slather* is compatible with the causative-plus-*with*-adjunct since the *target* can be construed as a type of patient, in that the entity which is slathered on can be construed as totally affected. The *with*-phrase is obligatory even though it is an adjunct, because the profiled status of the verb's *thick-mass* role requires that the role be expressed.

Verbs of the *heap-* and *cram*-classes are similar. Thus both full variants of the alternation are acceptable with *heap:*

(120) a.   Pat heaped mash potatoes onto her plate.
       b.   Pat heaped her plate with mash potatoes.

But again, none of the verb's roles may be left unexpressed:[16]

(121) a.   *Pat heaped mash potatoes.
       b.   *Pat heaped her plate.
       c.   *The mash potatoes heaped onto her plate.

The same is true of *cram:*

(122) a.   Pat crammed the pennies into the jar.
       b.   Pat crammed the jar with pennies.
(123) a.   *Pat crammed the pennies.
       b.   *Pat crammed the jar.
       c.   *The pennies crammed into the jar.

Thus, like the *slather*-class above, verbs of these classes must have three profiled participant roles:

heap ⟨**heaper, location, heaped-goods**⟩

cram ⟨**crammer, location, crammed-goods**⟩

The *load*-class of verbs also occurs in both constructions:

(124) a.   She loaded the wagon with the hay.
       b.   She loaded the hay onto the wagon.

It is not as clear in this case that all roles need be overtly expressed. While the agent and container roles are obligatory, as (125) shows, *load* can occur without an overtly expressed *theme* role, as in (126):

(125) a. *The hay loaded onto the truck.
     b.??Sam loaded the hay.

(126) Sam loaded the truck.

Can we still claim for this case that all roles are profiled? Notice that if the theme role is unexpressed, its filler must be presumed to be known to both speaker and hearer. That is, unless the context tells us *what* was loaded onto the truck, example (126) is infelicitous. Thus the theme role is allowed to be a *definite null complement* as discussed in section 2.4.5; this type of argument was argued to be lexically profiled despite the fact that it is not obligatorily expressed; it is indicated below by square brackets. The *load*-class therefore has three profiled roles as well:

load ⟨**loader, container, [loaded-theme]**⟩

Verbs of the *load*-class mark their theme roles as profiled, but optionally omissible if licensed by context to receive a definite interpretation.

Verbs of the *spray*-class, including *spray, splash, splatter, sprinkle,* work slightly differently. In the case of *splash,* both liquid and target roles must be expressed:

(127) a.   Chris splashed the water onto the floor.
     b.   Chris splashed the floor with water.
(128) a. *Chris splashed the water.
     b. *Chris splashed the floor.

Spray itself allows the liquid role to be unexpressed if it is given a definite interpretation:

(129) The skunk sprayed the car [ ].

In this example, the unexpressed liquid role is available to speaker and hearer through contextual given-ness; it is therefore still considered to be profiled.

Both these verbs can occur without an overtly expressed agent, as in (130):

(130) a.   Water splashed onto the lawn.
     b.   Water sprayed onto the lawn.

The following lexical entries capture these observations:

splash ⟨splasher, **target, liquid**⟩

spray ⟨sprayer, **target, [liquid]**⟩

The fact that the target can be construed as a type of patient, in that the entity which is sprayed can be construed as totally affected, allows *spray*'s roles to fuse with the argument roles of the causative construction. In particular, *spray* is licensed to occur in the caused-motion construction since the *sprayer* can be construed as a cause, the *liquid* as a type of theme, and the *target* as a type of goal-path. Similarly, the fact that the agent is not obligatory (i.e., non-profiled) allows *spray* to occur in the intransitive motion construction instantiated by (130).[17]

7.6 CONCLUSION

It has been argued that the argument structure associated with the interpretation of directly caused motion needs to be recognized as an English construction, independently of the lexical items which instantiate it. The evidence came from the fact that several aspects of the meaning of caused-motion expressions (causation, motion) and of their form (e.g., the direct object complement) are not generally predictable from lexical requirements or from other constructions.

The construction discussed in this chapter has as its basic sense a causer or agent directly causing a theme to move to a new location. The basic sense is extended in various ways, allowing the construction to appear with a variety of systematically related interpretations. As noted in section 3.3.2, this polysemy is strikingly similar to the polysemy argued to exist for the ditransitive construction in chapter 2.

In addition, specific semantic constraints have been proposed in an attempt to show principled patterns where there is apparent idiosyncrasy. These specific constraints can be interpreted as beginning to provide necessary conditions on the notion of "direct causation" (or of a "single event"). These constraints have been argued to involve a combination of lexical semantics and general world knowledge.

Finally, the *load/spray* alternation was discussed, and it was shown that both variants could be accounted for by understanding a single verb meaning to be able to fuse with two distinct constructions, the caused-motion construction and a causative construction plus *with*-adjunct.

8 The English Resultative Construction

8.1 Introduction

In this chapter the resultative construction, which was argued to be a metaphorical extension of the caused-motion construction (cf. chapter 3), is discussed in more detail. A great deal of attention has been focused recently on attempting to delimit the class of expressions to which resultatives can be applied (Bresnan and Zaenen 1990; Carrier & Randall 1992; Hoekstra 1988; Rappaport Hovav & Levin, 1991; Jackendoff 1990a; Levin & Rappaport 1990b; Napoli 1992; Randall 1983; Simpson 1983; Van Valin 1990a).

This chapter defends the position that the necessary constraint on the appearance of resultatives can be stated in semantic terms: the resultative can only apply to arguments that potentially (although not necessarily) undergo a change of state as a result of the action denoted by the verb. Such arguments are traditionally identified as *patients*. The traditional test for patienthood is that the expression can occur in the following frame (Lakoff 1976):

(1) a. What X did to ⟨patient⟩ was, . . .
     b. What happened to ⟨patient⟩ was, . . .

This idea would seem to be intuitive, given the fact that resultatives code a change of state caused by the verb. And in fact, this proposal has been approximated recently by Bresnan and Zaenen (1990), Jackendoff (1990a), and Van Valin (1990a). However, the existence of so-called "fake object" cases has been taken as an exception to the semantic constraint. Cases involving "fake objects," so named by Simpson (1983), include examples such as the following (the attested examples here and below come from Visser 1963):

(2) a. "Paulo, who had *roared* himself *hoarse,* was very willing to be silent." (OED: Mrs. Radcliff, Italian vii (1797))
     b. *He roared himself.

(3) a. "The Germans *cri'd* their throats *dry* with calling for a general Council." (OED: Leighton (1674) in Lauderdale Papers (1885))
     b. *The Germans cried their throats.

The postverbal NP in these cases is said to bear no semantic relation to the main verb, and therefore is viewed as being exceptional to the semantic constraint of patienthood. The existence of these cases has led several researchers to conclude that the phenomenon must be stated in syntactic terms (Simpson 1983; Carrier & Randall 1992; Rappaport Hovav & Levin 1991).

In what follows, I will continue to refer to these as "fake object" cases despite the fact that I will argue that the "fake object" should be treated as a semantic argument. In what follows, I restrict my attention to adjectival resultatives, although I intend the term "resultatives" to cover expressions which encode a resultant state with a PP as well.

The facts which must be accounted for are the following:

    1. Resultatives apply to direct objects of some transitive verbs:

(4)   "This nice man probably just wanted Mother to . . . kiss him unconscious." (D. Shields, *Dead Tongues* (1989))

(5)   "I had brushed my hair very smooth." (Ch. Brontë, *Jane Eyre* (1847))

(6)   "You killed it stone-dead." (W. Somerset Maugham, "Altogether" (1910))

    2. But they do not apply to direct objects of others:

(7)   *He watched the TV broken.

(8)   *He believed the idea powerful.

    3. Resultatives apply to subjects of passives which correspond to acceptable actives:

(9)   "I charged with them, and got knocked silly for my pains." (Rider Haggard, "King Solomon's Mines" (1889))

(10) The tools were wiped clean.

    4. They apply to the subjects of particular intransitive verbs, often associated with unaccusativity:

(11) The river froze solid.

(12) It broke apart.

    5. But they do not apply to the subjects of other intransitive verbs, often associated with unergativity:

(13) *He talked hoarse.

(14) *At his wedding, he smiled sore.

(15) *He coughed sick.

    6. Finally, as mentioned above, resultatives occasionally occur with so-called fake objects—that is, postverbal NPs that do not bear the normal argu-

ment relation to the matrix verbs. Some additional examples are the following (see also section 8.6):

(16) "Whose whole life is to *eat,* and *drink . . .* and *laugh* themselves *fat."* (OED: Trapp, Comm, and Epist. and Rev. (1647))

(17) "The dog would *bite* us all *mad."* (Dougl. Jerrold, Mrs. Caudle's Curt. Lect 4 (1846)) (This does not necessarily imply that the dog would bite us all)

(18) "She *laughed* herself *crooked."* (Benson, "Mr. Teddy" (1910))

8.2   THE STATUS OF THE POSTVERBAL NP

Following Simpson (1983), many researchers have assumed that the postverbal NP in the case of fake object resultatives is not an argument of the verb, whereas the postverbal NP of transitive resultatives is (Rappaport Hovav & Levin 1991; Bresnan & Zaenen 1990; Jackendoff 1990a; Napoli 1992). Carrier and Randall (1992) explicitly argue this point. They observe that some processes that are taken to apply only to direct internal arguments do not apply to fake object resultatives, although they do apply to regular resultative expressions. Specifically, they argue that middle formation, adjectival passive formation, and process nominalization apply to direct internal arguments. These processes are said to apply to "transitive" resultatives:

(19) Transitive Resultative: He hammered the metal (flat).
   a. Middle Formation: This metal hammers flat easily.
   b. Adjectival Passive: the hammered-flat metal
   c. Nominalization: the hammering of the metal flat

However, they do not apply to fake object resultatives:

(20) Fake Object Resultative: He drove his tires * (bald).
   a. Middle Formation: * Those tires drive bald easily.
   b. Adjectival Passive: * the driven-bald tires
   c. Nominalization: * the driving of the tires bald

Notice, though, that none of these constructions occurs across the board with all transitive resultatives either. For example, the following middles are based on transitive verbs, yet they pattern exactly like fake object cases in being ungrammatical:

(21) Middle Construction with transitive verbs
   a. *Pat kicks black and blue easily.
   b. *The washer loads full easily.
   c. *His face washes shiny clean easily.

And, as Jackendoff (1990a) has pointed out, most if not all adjectival passives and nominalizations based on transitive resultatives are also ungrammatical; this is exemplified in (22) and (23), respectively:

(22) Adjectival Passive of Transitive Verbs
  a. *the washed-shiny-clean face
  b. *the shot-dead man
  c. *the kicked-black-and-blue dog

(23) Nominalization of Transitive Verbs
  a. *the shooting of the man dead
  b. *the washing of the face shiny clean
  c. *the driving of him crazy

Notice we cannot claim that these facts provide evidence that even transitive resultatives do not have an internal argument since as (19) showed, some transitive resultatives *do* occur in these constructions. Examples with uncontroversial direct internal arguments differ on whether they can occur in these constructions. For example:

(24) Middle Construction
  a. This movie watches easily.
  b. *This movie sees easily.

(25) Adjectival Passive Construction
  a. the murdered man
  b. *the killed man (Lakoff 1965:46)

(26) Nominalization Construction
  a. the persuasion of people to new faiths
  b. *the persuasion of people to be quiet

Therefore, although there may well be an implication that if X occurs in the middle construction and adjectival passive construction and nominalization construction, then X is an argument, the converse is clearly false. So we cannot use these constructions to argue that fake object cases are *not* arguments. Neither Carrier and Randall nor the other researchers cited above provide other reasons for, or benefits to, attributing a non-argument status to the postverbal NP.

Each of the above constructions should be considered independently to see why some resultatives are compatible with them and others are not. To make this point, we will consider the case of middle formation in some detail below.

8.3 MIDDLE FORMATION

Middles require that the unexpressed agent argument be indefinite, interpreted as "people [or whatever the agent is] in general." Middles also require that the patient subject argument have a particular inherent quality which makes it primarily responsible for the property expressed in the predicate phrase (van Oosten 1977, 1984). Moreover, the unexpressed agent argument is typically interpreted as volitional, intending the result (if a result is entailed) as well as intending to perform the action denoted by the verb. To illustrate the fact that middles are normally interpreted as involving an (indefinite) volitional agent, notice the contrast between (27a, b) and between (28a, b):

(27) a.   This car drives with the greatest of ease.
      b. #This car drives with the greatest difficulty.

(28) a.   This wine drinks like it was water. (van Oosten 1977)
      b. #This wine drinks like it was vinegar.[1]

The same semantic features which are characteristic of middles can be captured by an appropriate paraphrase. For example, consider (29):

(29) The metal hammers flat easily.

This sentence is interpreted to mean:

(30) People can hammer the metal flat easily because of an inherent quality of the metal.

Several factors conspire to make fake object cases (as well as many transitive resultatives, and in fact, many simple transitives) incompatible with the middle construction. For one, fake object cases occur most readily with objects that are coreferential with the subject (Jackendoff 1990a); for example:

(31) He cried himself asleep.

(32) He talked himself blue in the face.

The restriction on middles that the unexpressed agent argument be indefinite suffices to rule the corresponding middles ungrammatical:

(33) *He cries asleep easily.

(34) *He talks blue in the face easily.

Another source of incompatibility stems from the constraint that the patient argument must be interpreted to have a particular inherent quality that makes the predicate true. Fake object cases, however, are often used as hyperbole to express the idea that the action performed was done to excess; in this use, it would be anomalous to attribute the predicate's holding to some particular property of the fake object referent. For example, consider (35):

(35) The joggers ran the pavement thin. (Carrier & Randall 1992)

This statement would not be used to describe an actual change in the thickness of the pavement, let alone to convey the idea that the pavement bore some kind of particular property which caused it to become thin from people running on it. Notice that the semantically analogous paraphrase (36c) is as unacceptable as the middle form (36b) itself:

(36)   a. #People can run that pavement thin easily because of an inherent qual-
              ity of the pavement.
        b. #That pavement runs thin easily.

Moreover, the fact that middles are typically used when the unexpressed indefi-
nite agent is understood to be volitional serves to render other possible fake object cases infelicitous. This is intuitively expected since fake object cases are often used to express a negative outcome; therefore assigning volitionality to the unexpressed agent results in anomaly.

Given the right context, we find that middles with fake objects can be greatly improved. For example, imagine that people in charge of props on a movie set are asked to drive fifty tires bald for a stunt. Insofar as speakers find (37) acceptable, the corresponding middle form (38) is also acceptable:

(37)   He drove fifty tires bald.

(38)   Go buy some cheap tires for that scene, those inexpensive tires drive bald
        really quickly.

Similarly, imagine that a farmer has had such trouble with stray dogs attacking his chickens that he breeds the chickens such that they wake up easily upon hearing any barking. In this context, insofar as speakers I have checked with accept (39), they also report (40) to be acceptable:

(39)   The dog barked the chickens awake.[2]

(40)   His chickens bark awake easily.

Thus, once closer attention is paid to the particular semantics associated with the middle construction, we can account for why fake object resultatives are not normally acceptable as middles, and we find that it is possible to concoct a context in which the semantics of a particular expression in fact *is* compatible with the middle construction.

8.4   OTHER ACCOUNTS

In the previous chapter (section 7.2), proposals by Pustejovsky (1991a), Rappaport Hovav and Levin (1991), and Hoekstra (1992) to compositionally derive the meaning of resultative expressions from their component parts were critically discussed. In this section, several additional accounts which are in some way related to the present proposal are reviewed.

8.4.1 Jackendoff 1990a

Jackendoff, although rejecting Carrier and Randall's specific arguments, also rejects the idea that the postverbal NP is an argument, instead suggesting that it is an adjunct. However, the postverbal NP fails traditional tests for adjunthood. For instance, it can appear as the subject of passives:

(41)   The baby was barked awake every morning by the neighbor's noisy dog.

It must occur directly after the verb, with no intervening material:

(42) *The dog barked ferociously the baby awake.

Omission of the postverbal NP results in a radical change of meaning. Also, only one postverbal NP can occur in a given clause:

(43) *The dog barked us them awake.

Therefore the claim that the postverbal NP of fake object cases is an adjunct and not an argument is unwarranted. In postulating an "adjunct rule" which can add the postverbal NP to the basic semantics of the verb, Jackendoff does, however, capture the basic insight that particular lexical items can be viewed as "fitting into" a construction with its own inherent semantics. A detailed comparison of Jackendoff's general approach and the one suggested here is presented in section 10.1.1.

8.4.2   Bresnan and Zaenen 1990

Bresnan and Zaenen (1990) argue that the critical factor is that the resultative be predicated of an argument that is intrinsically marked with the [−r] ('unrestricted') feature (cf. discussion in section 4.2.1). This feature is taken to be shared by subjects and direct objects, distinguishing them from prepositional objects and secondary objects. The [−r] feature is assigned as an intrinsic classification in either of two cases: (1) the argument bears the thematic role of patient (or "theme"), or (2) the argument is not assigned a thematic role by the verb.

The first case, the argument bearing the thematic role of patient, accounts for the majority of resultatives: those that are predicated of the direct objects of transitives, the subjects of passives, and the subjects of unaccusatives. The second case, that the argument is assigned no semantic role by the verb, is designed to account for fake object cases. Fake objects are assumed to be unsubcategorized for by the verb; therefore they are claimed to bear no thematic role, and are assigned the critical 'unrestricted' feature.

The problem with this account is that not only is the fake object not assigned a thematic role by the verb, it is not normally an argument of the verb, whether semantically empty or not; that is, it does not normally correspond to a *complement* of the verb. Bresnan and Zaenen fail to account for how it is that the internal object makes its way into the argument structure of the verb in order to receive its critical [− r] marking.

Both the approach and the problem with it can be stated more generally. Bresnan and Zaenen propose treating the verb with its fake object as a raising verb and, consequently, treating the postverbal NP as a complement but not as an argument. The question is how to account for the existence of this postverbal complement.

In order to deal with this issue, one could postulate a lexical rule which would add the internal argument to the argument structure of the verb (as has been done for applicatives in Alsina & Mchombo 1990) and for the resultative adjective itself within Bresnan and Zaenen's theory. And, if this were done, the additional argument could be assigned the thematic role of patient, since it is in all cases an affected argument. Providing thematic roles to arguments yielded by lexical rule is uncontroversial. This would allow Bresnan and Zaenen's account to reduce to the straightforward semantic constraint proposed here: resultatives can only be predicated of patient arguments.

8.4.3   Van Valin 1990a

The claim that resultatives can only be applied to patient arguments sounds on the face of it much like the account recently proposed by Van Valin (1990a). Van Valin argues that the resultative must be predicated of an "undergoer." However, he notes that "the label 'undergoer' should not be taken literally" (p. 226, fn. 6). In particular, undergoers do not correspond to patients in that it is not necessary that they potentially undergo a change of state; instead, in English any argument which can be passivized is taken to be an undergoer.[3] Therefore the undergoer condition is underconstrained with respect to resultatives, and would falsely predict examples which have undergoer but nonpatient arguments to be acceptable.

There is a more serious problem with Van Valin's account. He claims that "the constructions allowing resultative phrases are either accomplishments or achievements, all of which code a result state as part of their inherent meaning. Activity verbs, which are inherently atelic and therefore cannot in principle code a result state or have an undergoer argument, do not take resultative predicates" (p. 255). The problem stems from the fact that when Van Valin tries to exemplify the claim that resultatives only occur with accomplishments or achievements, he cites the *resulting* construction, not the construction before the resultative is added. For example, he notes that (44) is an accomplishment and that (55) is an achievement:

(44)   Terry wiped the table clean in/*for five minutes.

(45)   The river froze solid.

He then argues that unergative verbs do not allow resultatives, citing the following example:

(46)   He talked *in/for ten minutes.

However, this example is not parallel to the earlier examples since those already contained resultative phrases. There is no disagreement about the fact that expressions with a resultative are accomplishments or achievements, since the resultative phrase itself serves to bound the event. However, it is not the case that only such *independently* classifiable accomplishments or achievements occur with resultative predicates. For example, *push* in the following is an activity verb:

(47)   Terry pushed the door *in an hour/for an hour.

And yet *push* can occur with a resultative:

(48)   Terry pushed the door shut.

Also, *talk* is an activity verb, and yet the following is simply ungrammatical:

(49) *He talked himself.

In short, Van Valin's account begs the question of accounting for which predicates can occur with resultatives, and of how the fake object is related to the main verb.

8.5   THE EXISTENCE OF A RESULTATIVE CONSTRUCTION

The generalization we wish to capture is that the occurrence of resultatives can be predicted in purely semantic terms:

> Resultatives can only be applied to arguments which potentially undergo a change of state as a result of the action denoted by the verb.

As claimed above, the argument must therefore be classifiable as a type of *patient*. Again, we can use the traditional test for patienthood in order to determine whether the argument is of this sort.

(50)   a.   What X did to ⟨patient⟩ was, . . .
       b.   What happened to ⟨patient⟩ was, . . .

Notice there is no requirement that the predicate independently codes a change of state, only that it *potentially* causes a change of state.

By now, the reader who has been following along might guess how we can account for the occurrence of resultatives within a constructional approach.[4] A resultative construction is posited which exists independently of particular verbs that instantiate it. In order to account for fake object cases, we need to recognize that the construction itself can add a patient argument, besides adding the result argument to nonstative verbs which only have an "instigator" as profiled argument. Constructions as defined have semantics and are capable of bearing arguments. Thus the postverbal NP of the fake object cases is an argument *of the construction* although not necessarily of the main verb. Under this analysis, the verb retains its intrinsic semantic representation, while being integrated with the meaning directly associated with the construction. The resultative construction can be represented thus:

Resultative-Construction

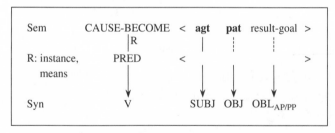

Figure 8.1

To see how the construction is able to add arguments, consider the following cases. Verbs such as *wipe* and *talk* can integrate into the resultative construction because they have compatible roles:

(51)   a.   wipe ⟨**wiper** wiped⟩
            He wiped the table clean.
       b.   talk ⟨**talker**⟩
            He talked himself blue in the face.

In general, for a verb to occur in a particular construction, the participant roles associated with the verb must fuse with the argument roles associated with the construction, according to the principles described in chapter 2. The participant roles of the verbs *talk* and *wipe* fuse with the argument roles of the construction as follows:

Composite Structure: Resultative + *wipe*

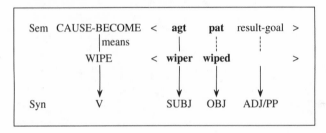

Composite Structure: Resultative + *talk*

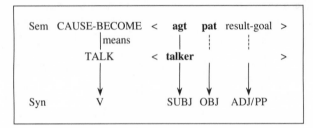

Figure 8.2

Thus the construction adds only the result-goal argument if the verb has a participant role which fuses with the patient argument of the construction, as is the case with *wipe*.[5] Alternatively, the construction can contribute both patient and result-goal roles, as is done in the case of *talk*.

Two other types of cases are ruled out. The construction itself does not prohibit a hypothetical verb with participant roles which are instances (types) of agent and result-goal from integrating into the construction, since the construction could presumably add the patient argument. However, the existence of such a verb is disallowed by the general constraint that instances of the result-goal role can only be predicated of patient-like roles.

A verb such as *become* with the participant roles '⟨**patient result-goal**⟩' cannot integrate with the construction, because the construction specifies that the agent role must be fused with an independently existing participant role of the verb (this is indicated by the solid line from the construction's agent role to the PRED role array).

Intransitive resultatives (i.e., resultatives with unaccusative verbs) require a slightly different construction; however, the more general constraint on patienthood is shared by two-place resultatives and intransitive resultatives (cf. examples 11 – 12):

Resultative-Construction

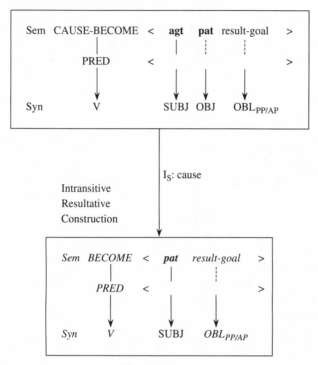

Figure 8.3

In figure 8.3 a subpart inheritance link relates the two-place resultative construction to the intransitive resultative construction. That two constructions are required is not necessarily a drawback of the present proposal. It seems that Italian allows only two-argument resultatives and does not allow resultatives with unaccusatives (cf. Napoli 1992).

The constructional approach captures the insight which led other researchers to explore the possibility that the postverbal NP is not an argument of the main verb, namely, that the postverbal NP does not intuitively correspond to any participant normally associated with the activity denoted by the main verb. The resultative construction is itself associated with a particular argument structure configuration, independently of verbs which instantiate it. Particular verbs retain their inherent semantics.

The analysis can motivate the existence of fake objects cases. Since the fake *reflexive* cases—the cases in which the resultative adjective is predicated of an argument which is coreferential with the subject—are the most common (according to Visser's survey), the most prototypical examples, and for some the only grammatical cases, we can understand fake object cases as having arisen from an expressive desire to predicate a change of state of an agent or instigator argument. A construction which adds a patient argument to the inherent argument structure of the verb allows the resultative to apply to a patient argument while allowing the patient argument to be corefential with the agent argument.

In addition, the syntactic expression of the postverbal NP would follow from general principles. Assuming a ternary branching structure (see Green 1973; Williams 1983; and Carrier & Randall 1992 for arguments against a small clause analysis) the patient argument is linked with OBJ by the canonical linking conventions of English (as suggested recently by, e.g., Gropen et al. 1991; Pinker 1989; Dowty 1991).

Further, an account which situates the possibility of resultative expressions in the semantics can naturally account for various semantic constraints on the construction. These are discussed in the following section.

Finally, this approach also allows us to capture the fact that there is a great deal of idiosyncrasy involved (Green 1972; Dowty 1979). Resultatives are often part of collocations with particular verbs. For example, *eat* is most colloquial with the resultative *sick:*

(52) a.   He ate himself sick.
      b.   ?He ate himself ill/nauseous/full.

Cry is most colloquial with the relative *to sleep:*

(53) a.   She cried herself to sleep.
      b.   ?She cried herself asleep.
      c.??She cried herself calm/wet.

The following minimal variants are markedly odd:

(54) ?He ate himself asleep.

(55) ?She cried herself sick.

What needs to be noted is that there are grammaticalized instances of the construction which are partially lexically filled.

Adopting a usage-based model of grammar as discussed in chapter 5 (which draws on the work of Bybee (1985) and Langacker (1987a)), novel extensions are acceptable to the degree that they conform to the semantic (and morphophonological) constraints on existing clusters of cases.

8.6   CONSTRAINTS ON THE RESULTATIVE CONSTRUCTION

The construction suggested above only provides a necessary condition on the appearance of resultatives. Several other, co-occurring constraints are required in order to begin to triangulate sufficient conditions on resultatives. In this section, the following restrictions will be argued to hold of (adjectival) resultative expressions generally (modulo cases in which the verb is lexically causative independently of the construction):

1. The two-argument resultative construction must have an (animate) instigator argument.

2. The action denoted by the verb must be interpreted as directly causing the change of state: no intermediary time intervals are possible.

3. The resultative adjective must denote the endpoint of a scale.

4. Resultative phrases cannot be headed by deverbal adjectives (Green 1972; Carrier & Randall 1992).

8.6.1   (Animate) Instigator Constraint

For many speakers (including myself), only animate instigator arguments are acceptable as subjects in two-argument resultative constructions. The animate argument is not necessarily an agent, since no volitionality is required:

(56)   She coughed herself sick.

(57)   She slept herself sober.

In some dialects, inanimate instigator arguments are also acceptable. For example:

(58)   The jackhammer pounded us deaf. (Randall 1983)

(59)   The alarm clock ticked the baby awake. (Randall 1983)

However, no speakers I checked with find instrument subjects acceptable:

(60) *The feather tickled her silly.

(61) *The hammer pounded the metal flat.

This constraint does not hold of lexical causatives, that is, verbs whose basic sense entails a change of state independently of the resultative:

(62)   Water filled the tub half full.

(63)   The sleeping pills made me sick.

8.6.2   Aspectual Constraint

There has been some disagreement about which aspectual classes can occur with resultative phrases. Van Valin (1990a) suggests that resultatives can only occur with telic predicates. Dowty (1979) and Jackendoff (1990a), on the other hand, suggest that resultatives can only occur with activity, or "un-bounded," predicates. It is at least generally agreed that resultatives cannot occur with stative verbs (Hoekstra 1988).

In this section I will argue that there is an aspectual constraint, but that it does not coincide with a distinction between telic and atelic predicates, both of which can be seen to appear in the resultative construction:

(64) a   Harry shot Sam dead.
      b. Harry shot Sam *for an hour. (telic, except on repetitive reading)
(65) a. Sam talked himself hoarse.
      b. Sam talked for an hour. (atelic)

The relevant constraint can be stated as follows:

> The change of state must occur simultaneously with the endpoint of the action denoted by the verb.

Allowed: Disallowed:

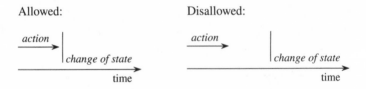

This constraint rules out cases in which there is any time delay between the action denoted by the verb and the subsequent change of state.

Notice that in a neutral context, *eat* with an unexpressed argument normally implies that the agent finished eating a meal:

(66) He (already) ate.

However, when *eat* occurs in the resultative construction, the eating is necessarily interpreted as extending over the period of time leading up to the change into a state of being sick. That is, (67) necessarily implies that the agent's continuous eating made him sick; it cannot imply that the meal he ate made him sick. Or consider (68):

(67) He ate himself sick.
(68) Sam cut himself free.

This sentence cannot be used to mean that Sam cut himself, causing his captors to release him in order to clean him up. It must mean that he cut whatever bonds were preventing him from being free, thereby immediately gaining his freedom. Similarly with (69):

(69)   Chris shot Pat dead.

This cannot be used to mean that Chris shot Pat and Pat later died in the hospital; instead it must mean that Pat died immediately from the shot.

This constraint can be interpreted as a consequence of a more general constraint that the causation must be direct: no intervening period is possible in a causal sequence (cf. discussion in 7.4.2. for other constraints on direct causation).

8.6.3   End-of-Scale Constraint

The type of adjective that can occur as a resultative is fairly limited. While adjectives "asleep/awake," "open/shut," "flat/straight/smooth," "free," "full/empty," "dead/alive," "sick," "hoarse," "sober," and "crazy" occur fairly regularly, others occur rarely if at all:

(70) *He drank himself funny/happy.

(71) *He wiped it damp/dirty. (Green 1972)

(72) *The bear growled us afraid.

(73) *He encouraged her confident.

(74) *He hammered the metal beautiful/safe/tubular. (Green 1972)

Most of the adjectives which can occur in the construction can be independently classified as having a clearly delimited lower bound and are therefore typically *nongradable* (Sapir 1944). Nongradable adjectives are said to be unable to appear (*ceteris paribus*) with quantifying phrases:

(75) ?a little sober

(76) ?a little flat/smooth

(77) ?a little alive/dead

(78) ?a little asleep/awake

(79) ?a little full/empty

(80) ?a little free

Intuitively, one cannot be a little sober, because one is either entirely sober or not sober: there is, all things being equal, no grey area.

Sick and *hoarse,* on the other hand, do not obviously code states with a clearly delimited lower bound:

(81)   a little sick

(82)   a little hoarse

However, when used in the fake object construction, they are interpreted as delimiting the clear boundary beyond which the activity cannot continue. Consider (83, 84):

(83) He ate himself sick.

(84) He talked himself hoarse.

These expressions imply that the agent ate to the point where he could eat no more, or talked to the point where he could talk no more. Notice how in this context the adjectives receive a nongradable interpretation:

(85) ?He talked himself a little hoarse.

(86) ?She ate herself a little sick.

The adjectives *crazy* and *silly* are similar in this respect:

(87) He drove her crazy/bananas/bonkers/mad/insane.

(88) He tickled her silly.

They imply that the patient argument has "gone over the edge," beyond the point where normal functioning is possible (of course they are typically used as hyperbole, not literally).[6] *Render* is interesting in that it lexicalizes this constraint, requiring a resultative adjective which codes a state of loss of function (that is, the property must be the negative end of a scale):

(89) a.   It rendered them speechless/impotent/obsolete.
      b.??It rendered them alive/full/free.

Therefore it is fair to say that the resultative of the fake object construction codes a clearly delimited endpoint.[7] The endpoint may be on some absolute scale (in the case of inherently nongradable adjectives) or on a scale of functionality, in which case continued functioning is impossible beyond it:

Exceptions to this generalization are of two kinds. First, there are verbs which are lexically causative, independently of the resultative construction. These verbs are much freer in the semantic and syntactic type of resultative phrase they may occur in than *render* and productive cases:

(90) a.   He made the metal safe/pretty/tubular/damp/dirty.
      b.   He made her a queen.

(91) a. He painted his house pink.
      b. He painted his house a bright shade of red.

Other exceptions to the above generalization have been attested, but aside from their apparent rarity, each can be seen to have a distinctly novel character. In general, exceptional cases tend to be from the same semantic domain as more conventionalized cases, and can be seen as one-shot novel extensions from a grammaticalized pattern:

(92) "Bees will suck themselves tipsy upon varieties like the sops-of-wine." (OED: J. Burroughs, "Locusts and Wild Honey" (1879))

(93) Till he had drunk himself sleepy. (R. L. Stevenson, *Treasure Island* (1893))

There is one attested case, though cited by Rappaport Hovav and Levin (1991), which truly seems to fly in the face of this generalization:

(94) "Look, isn't it lovely? It's the stale loaf I put out for the birds and they've pecked it really pretty." (cited by Rappaport Hovav and Levin 1991; from Z. Wicomb, *You Can't Get Lost in Cape Town* (1987))

However, this example is judged ungrammatical by American English speakers I have asked. It is possible that South African English does not have the end-of-scale restriction.

8.6.4   Restriction against Deverbal Adjectives

A general constraint that is widely recognized is that resultatives cannot be adjectives derived from either present or past participles (Green 1972; Carrier and Randall 1992):

(95) a. She painted the house red.
      b. *She painted the house reddened.
      c. *She painted the house reddening.
(96) a. She shot him dead.
      b. *She shot him killed.
      b. *She shot him dying.
(97) a. She kicked the door open.
      b. *She kicked the door opened.
      c. *She kicked the door opening.

This restriction has been attributed to a semantic clash of aspect (Carrier & Randall 1992); however, the exact nature of the cause clash—clash has proved elusive.

8.7 CONCLUSION

This chapter has argued that the semantic restriction that resultatives can only apply to patient arguments is viable, even in the case of fake object resultative expressions, despite recent arguments to the contrary. This analysis has the advantages of (1) assimilating fake object cases to other transitive resultative cases, (2) motivating the existence of fake object cases, (3) allowing for the existing idiosyncrasy in a natural way, (4) predicting the syntax of the construction from canonical linking patterns and without ad hoc stipulations, and (5) accounting for semantic constraints in a natural way.

The following specific semantic constraints were proposed in order to restrict the applicability of the lexical rule (or the instantiation of the construction): (1) two-argument resultatives must have an instigator argument; (2) the causation involved must be direct, with no intervening time periods allowed; (3) the resultative adjective must have a clearly delimited lower bound; and (4) the resultative adjective must be considered a type of path phrase, which accounts for several co-occurrence restrictions.

9 The *Way* Construction

9.1 INTRODUCTION

The construction to be discussed in this chapter can be skeletally represented as follows (where V is a nonstative verb, and OBL codes a directional):

[SUBJ$_i$ [V [*POSS$_i$ way*] OBL]]

Several large corpora were searched for examples. The majority of the examples (1,050 out of 1,177) are from the Oxford University Press corpus (oup), which has been cited in earlier chapters already.[1] Additional examples have been culled from exhaustive searches of the Wall Street Journal 1989 corpus (wsj), the Lund corpus consisting of various spoken dialogs (lund), and the United States Department of Agriculture corpus (usda).

9.2 THE EXISTENCE OF THE CONSTRUCTION

Instances of this construction imply that the subject referent moves along the path designated by the prepositional phrase. The construction's semantics cannot be fully predicted on the basis of the constituent parts of the construction. For example, (1) entails that Frank moved through the created path out of the prison.

(1) Frank dug his way out of the prison.

Similarly, (2) entails that Frank managed to travel to New York.

(2) Frank found his way to New York.

However, none of the lexical items involved entails motion. To see this, compare (1) and (2) with (3) and (4) below:

(3) Frank dug his escape route out of prison.
(4) Frank found a way to New York.

The only interpretation for these examples is one in which the prepositional phrase modifies the direct object. Neither (3) nor (4) entails motion:

(4') Frank dug his escape route out of prison, but he hasn't gone yet.
(5') Frank found a way to New York, but he hasn't gone yet.

This is in contrast with examples (1) and (2), which do entail motion:

(2′) *Frank dug his way out of prison, but he hasn't gone yet.

(3′) *Frank found his way to New York, but he hasn't gone yet.

The only difference between (1) and (3) is that *way* is replaced by *escape route.* Example (4) prevents us from postulating that *way* codes motion, because *way* is present in this example and yet the sentence does not entail motion. Without belaboring the point, it should be pointed out that motion is not dictated by the combination of bound pronoun and *way* either, since the expression in (5) does not entail motion:

(5)   He knows his way around town.

Here entailment of motion is not present because the verb *know* is stative, and the construction requires a nonstative verb.

Salkoff (1988) and Jackendoff (1990a) also point out that this construction provides evidence for the claim that verbs do not exclusively determine complement configuration. One solution Jackendoff proposes is that examples such as those in (1–2) instantiate a particular clause-level construction: a pairing of form and meaning that exists independently of the particular verbs which instantiate it. As he suggests, "in a sense, the *way*-construction can be thought of as a kind of 'constructional idiom,' a specialized syntactic form with an idiomatic meaning, marked by the noun *way*" (1990a:221).

Levin and Rapoport (1988) suggest instead that each verb in the construction has a special motion sense, perhaps generated by a lexical rule, which predicts its occurrence in this pattern. However, this pattern occurs with an enormous variety of verbs. For example, we would need to posit such a motion sense for each of the verbs in the following:

(6)   a.   ". . . he'd *bludgeoned* his way through, right on the stroke of half-
              time." (oup)
       b.   "[the players will] *maul* their way up the middle of the field." (oup)
       c.   ". . . glaciers which had repeatedly *nudged* their way between En-
              gland and Wales." (oup)

That is, we would need a special sense of *bludgeon,* 'to move by bludgeoning,' a special sense of *maul,* 'to move by mauling,' and so forth. These senses are intuitively implausible. The following examples (presented in section 5.5.2 and repeated below) involving metaphorical motion would be even more difficult to imagine as projections from a lexical subcategorization:

(7) a. ". . . their customers *snorted and injected* their way to oblivion and sometimes died on the stairs." (oup)
     b. "But he consummately *ad-libbed* his way through a largely secret press meeting." (oup)
     c. "I cannot inhabit his mind nor even *imagine* my way through the dark labyrinth of its distortion." (oup)
     d. "Lord King craftily *joked and blustered* his way out of trouble at the meeting." (oup)

If new senses *were* involved, then it would follow that each of the verbs above would be ambiguous between its basic sense and its sense in this syntactic pattern. Therefore we would expect that there would be some language that would differentiate the two senses by having two independent verb stems. However, to my knowledge there is no language that has distinct verb stems for any of the meanings that would be required for the examples in (6–7).

In addition to being implausible, positing additional verb senses can be seen to be less parsimonious than associating the semantic interpretation directly to the construction. The reason for this stems from the fact that the proposed senses occur only in this construction; they are not available when these verbs are used with other valences:

(8) *Chris bludgeoned/mauled/snorted and injected into the room.

The same is not true of verbs which clearly do lexically code literal or metaphorical motion, for example, *inch* and *worm:*

(9) a. Lucky may have inched ahead of Black Stallion.
     b. He can't worm out of that situation.

That is, both *inch* and *worm* can be used as (metaphorical) motion verbs even when they are not used specifically in the *way* construction.

Therefore, not only would we need to stipulate the existence of additional senses for each of the verbs in examples (6–7), but we would have to further stipulate the fact that the new verb senses can only occur in this particular syntactic configuration. Clearly it is more parsimonious to attribute the motion interpretation directly to the construction itself.

Given that the interpretation of *way* expressions is not fully predictable from the semantics of the particular lexical items, a constructional analysis will be adopted here. An explicit statement of the construction will be preceded by a more specific analysis of the construction's semantics, since it will be argued that the syntax of the construction is motivated by its semantics.

9.3   THE SEMANTICS OF THE *WAY* CONSTRUCTION

9.3.1 Two Different Senses

Both Levin and Rapoport (1988) and Jackendoff (1990a) suggest two distinct paraphrases of this construction, one in which the verb designates the means of motion, the other in which the verb designates some other coextensive action or manner.[2] For example, Jackendoff notes that (10) is interpretable in either of the two ways given in (11):

(10) Sam joked his way into the meeting.

(11) a.   Sam got into the meeting by joking. (means)
      b.   Sam went into the meeting (while) joking. (manner)

These paraphrases together are taken to comprise a disjunctive interpretation. However, logical disjunction is completely symmetric, and there are reasons to think that the means and manner interpretations do not have equal status in the grammar—that in fact the means interpretation is primary.

In the Oxford University Press, Lund, Wall Street Journal, and Department of Agriculture corpora, verbs which designated a coextensive action or manner, as opposed to the means of motion, were rare. In fact, the total number of occurrences of verbs with manner interpretation in the corpora was 40 out of 1,177, or less than 4% of the data (the percentage was no more than 4% for any of the corpora).

In addition, it is argued below that the syntactic form of the construction is motivated by the semantics associated with the means interpretation. The same cannot be said of the manner interpretation. That is, recognizing a cline of analyzability (or compositionality) of idiomatic expressions (cf. Nunberg, Wasow & Sag 1982; Gibbs 1990), the means interpretation will be argued to be more analyzable than the manner interpretation.

Finally, not all speakers find the strict manner interpretation acceptable. A case in point is (12), one of Jackendoff's examples:

(12) He belched his way out of the restaurant.

When asked for judgments of this sentence, which was intended to have a manner interpretation (the subject went out of the restaurant while belching), several speakers I checked with concocted situations in which the belching instead was the *means* by which motion was achieved. For example, one speaker suggested that the sentence would be acceptable in the context that the other diners found the belching so objectionable that they cleared a path through which the offending party could exit. Another speaker suggested that the sentence would be acceptable if the belching were understood to be a means of propulsion. Others, including myself, find the manner interpretation only marginal.

Since as we've seen it is natural for constructions to be associated with a central sense, and with extensions from that sense, these facts can be easily accounted for. We can analyze the manner interpretation as an extension of the more basic means interpretation. This analysis predicts, for example, that there are no speakers who accept only the manner interpretation and not the means interpretation. And to date, I have found none.

Interestingly, there is diachronic evidence that the means interpretation of the construction predates the manner interpretation by more than four centuries. The first citation of this pattern in the OED is from the year 1400: *"I made my way . . . unto Rome."* [3] The first citation with any other verb is from 1694: *"[He] hew'd out his way by the power of the Sword."* The first example cited in the OED that involves a pure manner interpretation, *"The muffin-boy rings his way down the little street,"* is dated 1836, more than a century after the construction was first used productively with a means interpretation, and more than four centuries after the first citation with *make*. The diachronic data of course do not directly support the claim that the means interpretation is synchronically more basic; however, they do provide evidence that the extension from means to manner is a reasonable move for speakers to make, since at least one generation of speakers was willing to extend the pattern in just this way.

To summarize, it has been argued in this section that the means interpretation is the more central, or *basic,* interpretation of the construction. The manner interpretation has been argued to be a less basic extension, on the grounds that (1) manner examples were rare in each of the four corpora analyzed (accounting for less than 4% of the data), (2) speakers' judgments as to the acceptability of the manner cases range from unacceptable to marginal to acceptable, while the means cases are all fully acceptable, and (3) the means interpretation diachronically preceded the manner interpretation by several centuries. A fourth reason for claiming that the means interpretation is more basic follows from the observation, detailed below, that the syntactic form of the construction can be motivated by the means interpretation but not by the manner interpretation. In the following, a particular semantic constraint on the means interpretation is proposed, namely, that the motion must be through a literal or metaphorical self-created path. This constraint is argued to play a crucial role in motivating the syntactic form of the construction.

9.3.2   The Means Interpretation: Creation of a Path

Jespersen (1949) had the basic insight that the direct object, POSS *way,* was a type of "object of result." This can be interpreted to mean that the path (the *way*) through which motion takes place is not preestablished, but rather is *created* by some action of the subject referent. This observation can be used to account for the fact that, with the exception of the pure manner interpretation, the construction is used to convey that the subject moves despite some external difficulty, or in some indirect way: the path is not already established, but must in some sense be created by the mover. Consider the following:

(13) Sally made her way into the ballroom.

This sentence is understood to imply that Sally moved through a crowd or other obstacles. It cannot be used to mean that Sally simply walked into an empty ballroom. In the case of metaphorical motion, the necessity of creating a path implies that there is some difficulty or metaphorical barrier involved. For example, notice also the difference in acceptability between the following:

(14) a.??Sally drank her way through the glass of lemonade.
      b. Sally drank her way through a case of vodka.

Example (14b) is much more acceptable because it is much easier to construe drinking a case of vodka as requiring that some barrier be overcome than drinking a glass of lemonade.

In fact, the most common interpretation of this construction involves motion through a crowd, mass, obstacle, or other difficulty—that is, there is some reason why a path needs to be created. The verb either lexically subcategorizes for the construction (e.g., *make*) or designates the means by which the motion is achieved. For example:

(15) "For the record, Mr. Klein, as lead climber for the Journal team, pushed his way past the others, trampling the lunch of two hikers in his black army boots, and won the race to the summit." (wsj)

(16) "In some cases, passengers tried to fight their way through smoke-choked hallways to get back to their cabins to get their safety jackets." (wsj)

(17) "For hours, troops have been shooting their way through angry, unarmed mobs." (wsj)

Contain verbs, such as *thread, wend, weave,* encode a slightly different interpretation. They involve deliberate, careful, methodical, or winding motion. In these cases as well as in the cases which involve some external difficulty, the subject is not moving along a preestablished path. For example:

(18) "This time, with no need to thread his way out, he simply left by the side door for a three-day outing." (wsj)

(19) "A couple in fashionable spandex warm-up suits jogs by, headphones jauntily in place, weaving their way along a street of fractured and fallen houses." (wsj)

The fact that the construction entails that a path is created to effect motion—that the motion takes place despite some kind of external difficulty or is winding and indirect—accounts for why high-frequency, monomorphemic (basic or superordinate level) motion verbs are typically unacceptable in this construction:

(20) *She went/walked/ran her way to New York (Napoli cited by Jackendoff 1990a)

(21) *She stepped/moved her way to New York.

These vanilla motion verbs do not normally imply that there is any difficulty or indirect motion involved, an implication which is required by the means interpretation of the construction. (Note that the manner interpretation is also unavailable, since these verbs do not code any salient manner.) If a context is provided in which a basic-level motion verb is understood to imply motion despite difficulty, these cases are decidedly better:

(22)   a.   The novice skier walked her way down the ski slope.
       b.   The old man walked his way across the country to earn money for charity.

Another case in which a (metaphorical) path may need to be created is given if there are social obstacles standing in the way. Contrast the following examples:

(23)   a. #Welcome our new daughter-in-law, who just married her way into our family.
       b.   Welcome our new daughter-in-law, who just married into our family.

Example (23a) is pragmatically odd because it implies that the daughter-in-law in question managed to get herself into the family by marriage, and such an implication is incongruent with a sincere welcome. The following example is relevantly similar:

(24)   Joe bought his way into the exclusive country club.

This example entails that Joe managed to get himself into the country club despite social obstacles. The necessity of the metaphorical creation of one's own path despite social obstacles can account for the implication that the subject referent used some unsanctioned means to attain his goal. That is, if there are social obstacles preventing one from attaining a goal, the only way to attain the goal is to violate the social constraints. Attested examples of this class include *bribe, bluff, crapshoot, wheedle, talk, trick, con, nose, sneak, weasel, cajole, inveigle*. Several lexical items seem to lexicalize this sense, for example, *worm, weasel,* and *wrangle*.

The claim then is that *way* is analyzable as a literal or metaphorical path that is created by the action denoted by the verb. This accounts for the semantic constraint that the motion is effected despite some external physical or social obstacles, by forging a path through or around those external obstacles.

Support for the claim that *way* is analyzed as a meaningful element comes from the fact that it can appear with modifiers. The following examples are attested:

(25) a. ". . . the goats wending their familiar way across the graveyard . . ." (oup)

     b. "[He] decided from then onwards that he could make his own way to school . . ." (oup)

In example (25a), *familiar* is a modifier of *way*—that is, the path is familiar. Similarly, in (25b), *way* is internally modified by *own*. These facts argue that the phrase POSS *way* is not an arbitrary syntactic tag of the construction, but rather plays a role in the semantics of the construction.

Further support for the claim that the construction at least historically was associated with the creation of a path comes from the fact noted above that the verb *make*, a verb which normally means "create," has had a privileged status with respect to this construction: this verb was used in the construction for almost three centuries before the construction was extended to be used with other verbs, according to citations in the OED.

Make continues to be closely associated with the construction insofar as it is used with greater frequency than any other single verb, accounting for 20% of the tokens. This suggests that *make* may well have a privileged status synchronically as well.

Finally, the recognition that the *way* is an effected entity motivates the syntactic form of the construction. As stated at the onset of this chapter, Jackendoff notes that there are reasons to assign the construction the structure:

$$[\text{SUBJ}_i \, [\text{V} \, [POSS_i \, way] \, \text{OBL}]]$$

He argues that the noun phrase 'POSS *way*' is a direct object, rather than some kind of syntactic adjunct or measure phrase, because nothing can intervene between the V and this phrase:

(26) *Bill belched noisily his way out of the restaurant. (1990a:212)

The OBL phrase coding the path is argued to be a sister of the verb, rather than a modifier of *way*, on the grounds that an adverb may intervene between the two complements, indicating a constituency break:

(27) a. Bill belched his way noisily out of the restaurant. (1990a:212)
      b. "He made his way cautiously along the path beside the lake." (oup)

Given the semantics of the means interpretation described above, the construction can be viewed as a kind of conventionalized amalgam that combines the
syntax and semantics of creation expressions such as (28), which have two
arguments—a creator and a "createe-way"—with the intransitive motion construction exemplified by (29), which has two arguments—a mover (theme) and
a path.

(28) He made a path.
(29) He moved into the room.

The *way* construction syntactically and semantically amalgamates these two
constructions into a structure with three complements: the creator-theme, the
createe-way, and the path. Thus the *way* construction can be viewed as inher-
iting aspects of both the creation and the motion constructions, while nevertheless existing as an independent construction in its own right.

   The semantics involves both the creation of a path and movement along that
path. As was true for the constructions discussed in previous chapters, the verb
may, but need not necessarily, code the semantics associated with the construction directly. Cases in which the verb does directly code the semantics of the
construction include *worm, inch,* and *work.* In other cases, the verb may designate the means of effecting the action designated by the construction; that is
to say, the verb may code the means of effecting motion through a self-created
path. This is represented below by the means link between PRED, representing
the verb sense, and the CREATE-MOVE predicate.

Way Construction: Means Interpretation

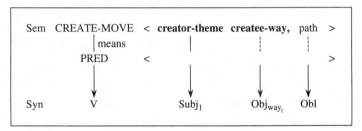

Figure 9.1

   Any argument that the verb obligatorily expresses must be fused with one of
the arguments associated with the construction. For example, the verb *push* has

one obligatory argument, the pusher. This argument is fused with the creator-theme argument of the construction (a pusher can be construed as a type of creator-theme). Both the createe-way and the path phrase are contributed by the construction. The fused composite structure is represented below.

Composite Structure: *Way* Construction + *push*

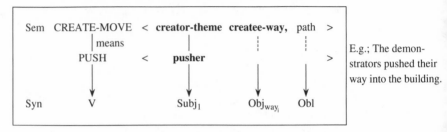

Figure 9.2

The verb *lurch* has two obligatory arguments, a lurcher and a path. These two arguments are fused with the creator-theme and path arguments of the construction, respectively. In this case the direct object argument, the createe-way, is contributed by the construction:

Composite Structure: *Way* Construction + *lurch*

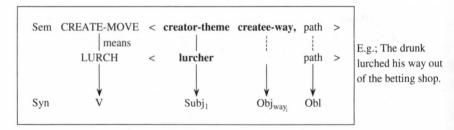

Figure 9.3

The syntactic form of the construction, given above, does not actually require much stipulation. The fact that the construction takes the syntactic form it does is strongly motivated by general principles. That is, the *POSS-way* phrase is linked to the direct object, because effected entities are generally direct objects. The fact that the path argument is linked to an adverbial directional follows from the fact that it is coding a path. The fact that the creator-theme is linked to the subject follows from the fact that creators and self-propelling themes are generally linked to the subject. It is only necessary to state that the created-way argument must be realized in a particular fixed way: by the bound pronoun plus *way.* Thus, little needs to be said about the syntax of the construction once its special semantic properties are captured.

To summarize, the recognition that the path of motion is not preestablished and must be created by the mover accounts for the fact that the means interpretation of the construction always entails that the subject referent moves despite external difficulty or in some indirect way. This observation allows us to analyze *way* as a meaningful element, designating the path of motion. It also allows us to account for the fact that *make,* a verb which in its basic sense means "create," has a privileged status both diachronically (in being the first verb to be used in the construction) and synchronically (in being the most frequently used verb in this construction). Finally, recognition that *way* designates an effected entity allows us to motivate the syntactic form of the construction.

9.3.3   The Manner Interpretation

The following attested examples do not admit of a means interpretation:

(30) "[They were] clanging their way up and down the narrow streets . . ." (oup)

(31) ". . . the commuters clacking their way back in the twilight towards . . ." (oup)

(32) "She climbed the stairs to get it, crunched her way across the glass-strewn room . . ." (oup)

(33) "He seemed to be whistling his way along." (oup)

(34) ". . . he was scowling his way along the fiction shelves in pursuit of a book." (oup)

For example, (30) does not entail that the clanging was the *means* of the motion, only that it was the co-occurring manner. Again, not all speakers accept this type of example, but many do, at least marginally.

While many of the attested manner cases involve motion in the face of some external difficulty or obstacles, just like the means cases, this does not seem to be a general constraint on the interpretation of the manner cases. Examples (31) and (33), for instance, do not imply any external difficulties. It is because of this that the syntax is claimed to be less analyzable in the case of the manner interpretation: there is no necessary implication that a path must be created. The manner interpretation only entails that the subject referent moves along a (possibly pre-established) path. Thus the *way*-phrase in direct object position is not motivated. It is predicted that internal modification of *way* in the manner interpretation should be less acceptable than in the means interpretation. And in fact, this prediction seems to be borne out:

(35)   a.??Joe whistled his own way to the street. (manner)
     b.   Joe dug his own way to the street. (means)

The interpretation associated with this sense can be represented as follows:

Way Construction: Manner Interpretation

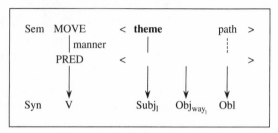

Figure 9.4

In the representation above, the construction designates a two-place relation; the *way*-phrase is not represented in the semantics of the construction, but is instead encoded as a syntactic stipulation about the form of the direct object complement.

9.3.4   Constructional Polysemy

It has been argued that the manner interpretation is an extension of the more basic means interpretation. The relation between the two senses can be represented thus:

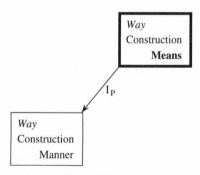

Figure 9.5

The constructions are related by an I_P (polysemy inheritance) link (cf. discussion in section 3.3.2). The arrow specifies that there is a systematic relationship between the two senses, in that the form is inherited from the means case by the manner case.

The claim that this is a case of constructional polysemy raises a question: what is the relation between means and manner that allows manner to become an extension?

Note that the means and manner of a motion event are often coextensive, in that the means of motion often determines the manner of motion. For example, consider the use of *roll* in (36):

(36)   Joe rolled the ball into the room.

In this example the verb specifies not only the means of motion (Joe caused the ball to move by rolling) but also the manner of motion (Joe caused the ball to move while rolling). The same is true of many motion verbs, e.g., *float, wiggle, jump, skip*. It is likely that the polysemy was created when speakers began to decouple these two often co-occurrent features. This type of decoupling is in fact a known source of *lexical* polysemy (cf. Traugott 1989).

In order to further motivate the link between means and manner, note that this construction does not provide the only instance of this particular polysemy in the language. English *with* is used both as an instrumental or means marker and as a marker of manner:[4]

(37)   Bob cut the bread with a knife. (means)
(38)   Bob cut the bread with care. (manner)

How is similarly polysemous:

(39)   How did you cut the turkey? Answer:
       a.   With a knife. (means)
       b.   Carefully. (manner)

Interestingly, the noun *way* itself is polysemous between means and manner senses. An example in which *way* is used to mean "means" is (40):

(40)   Pat found a way to solve the problem.

Way is used to mean "manner" in (41):

(41)   He had a pleasant way about him.

It is possible that this latter fact has encouraged the use of this particular construction with a strict manner interpretation, although it is somewhat unclear how to state this parallelism between the polysemy of *way* and the polysemy of the construction.

To summarize, attested examples involving a manner interpretation—that is, a purely co-extensive activity not causally related to the motion—while rare, do exist. These cases do not necessarily involve motion in the face of obstacles or indirect motion. Therefore there is no reason to think that these examples indicate the creation of a path, and thus the syntactic expression of *POSS way* is not motivated directly by the semantics of the manner extension. Instead, the manner sense of the construction was argued to be an extension of the more basic means sense. The link between means and manner was motivated by noting several instances of similar lexical polysemy, including the polysemy of the noun *way* itself.

9.4   SEMANTIC CONSTRAINTS
9.4.1 Unbounded Activity

There is a constraint that the verb necessarily designate a repeated action or unbounded activity (Jackendoff 1990a):

(42)   a.   Firing wildly, Jones shot his way through the crowd.
      b.   *With a single bullet, Jones shot his way through the crowd.
(43)   a.   Bill punched his way through the crowd by pummeling everyone in his path.
      b.   *Bill punched his way through the crowd by leveling the largest man and having everyone else step aside.

For the same reason, we find the following to be unacceptable:[5]

(44) *She dove her way into the fire.

(45) *She jumped her way over the ditch.

This constraint also seems to hold of the manner interpretation. For example, consider (46):

(46)   He hiccupped his way out of the room.

This sentence entails that there were a series of hiccups occurring over time, rather than a single hiccup.

9.4.2   Self-propelled Motion

A further constraint on the means interpretation is that the motion must be self-propelled:[6]

(47) *The wood burns its way to the ground.

(48) *The butter melted its way off the turkey.

This constraint serves to rule out the use in the construction of verbs which are commonly classified as unaccusative (Perlmutter 1978; Burzio 1986), since unaccusativity has been argued to be correlated with lack of agentivity or lack of self-initiation (Van Valin 1990a; Levin & Rappaport Hovav 1990a; Zaenen,

1993). However, it seems that the relevant constraint is semantic, insofar as the normally unaccusative verbs *grow* and *shrink* are attested in the data, with an agentive interpretation:

(49) "The planned purchase furthers Bull's strategy of trying to grow its way out of its extensive computer-marketing problems." (wsj)

(50) "The bank-debt restructuring is the centerpiece of Lomas Financial's months-long effort to shrink its way back to profitability after two straight years of heavy losses." (wsj)

The subject referent need not be volitional, or even human, as long as the motion is construed as self-propelled:[7]

(51) ". . . sometimes it [the cyst] forces its way out of the ((plumpton)) at the top." (usda)

(52) "The large seeds sprout quickly and dependably and the strong seedlings can push their way through crusted soil." (usda)

There are two lexical exceptions to the constraint that the motion must be self-propelled, *work* and *find:*

(53) "The spending bills working their way through Congress don't present much of a problem in terms of the Gramm-Rudman law." (wsj)

(54) "Bolivia estimated that about half its sacred textiles had been smuggled out of Bolivia and had found their way into American collections." (wsj)

Find in this use is further distinguished from the general case in that only the goal or endpoint of the path can be made explicit—the route itself may not be expressed.[8] This is evident from the fact that examples with an explicit path are unacceptable:

(55) *The textiles found their way through customs.

(56) *The statements found their way toward the right people.

The constraint that the motion must be self-propelled does not seem to hold of the manner interpretation; the following manner example, which does not involve self-propelled motion, is attested:

(57) " 'I knitted my way across the Atlantic,' he reveals." (oup)

However, even in this case, the action designated by the verb—the knitting—is performed agentively. Speakers only accept manner examples in which the action designated by the verb is self-initiated (otherwise such speakers would be able to give examples (47) and (48) a manner interpretation, and they do not).

9.4.3   Directed Motion

Related to the constraint that the motion must be self-propelled is the fact that the motion must be directed—it cannot be aimless. This accounts for the unacceptability of the following examples:

(58) *She wandered her way over the field.

(59) *She meandered her way through the crowds.

Notice that it is actually not possible to state the constraint as a constraint on a class of *verbs* per se because the constraint applies to nondirected motion expressed by means of prepositions such as *among* as well:

(60) *Joe shoved his way among the crowd.
      (Joe shoved his way through the crowd.)

This constraint also does not strictly hold of the manner interpretation, although there may be a tendency to prefer directed motion. Dialects which allow the pure manner interpretation differ as to the acceptability of (58–60), with some accepting them fully and others marginally.

9.5   THE LEXICAL COMPLEX PREDICATE APPROACH

In addition to suggesting a constructional analysis of the *way* expressions, Jackendoff (1990a) proposes an alternative solution. He does not decide between the two proposals. The alternative proposal is that verbs which appear in the construction undergo a lexical rule, turning, for example, *push* into *push POSS way.* The complex predicate *push POSS way* can then be argued to select for a path argument.

The problem with this proposal is that *POSS way* is not only an N but an NP. Therefore the complex predicate analysis would have to posit a maximal phrase internal to a word. Recent incorporation proposals (e.g., Baker 1988) have allowed incorporated proforms, but not full NPs complete with determiners and optional modification.

One might argue that the NP is unanalyzed, and is therefore simply a string that forms part of the complex predicate. Support for this idea might be drawn from the fact that, as Jackendoff notes, modifiers of *way* often have only an external (adverbial) interpretation:

(61) a.   Bill joked his insidious way into the meeting. (Jackendoff 1990a)
         (Meaning: Insidiously, Bill joked his way into the meeting.)
    b.   "[They] make their noisy way along the Rue Saint Antoine." (oup)
         (Meaning: They noisily made their way along the Rue Saint Antoine.)

c.   "They made their weary way home." (oup)
      (Meaning: They wearily made their way home.)

However, there is reason to think that the phrase *POSS way* is analyzed syntactically and semantically as a noun phrase. First, it has the normal internal structure of a noun phrase, and is not idiosyncratic syntactically in any way. Also, the possessive phrase is controlled by the subject, and thus its realization is not predetermined.

The fact that internal modification of *way* is possible, as in examples (25a, b), repeated here as (62a, b), further supports the claim that the noun *way* is meaningful, since being meaningful is a prerequisite to being available for modification:

(62) a.   ". . . the goats wending their familiar way across the graveyard . . ." (oup)
      b.   "[He] decided from then onwards that he could make his own way to school . . ." (oup)

That is, if *way* were simply a semantically empty syntactic marker of the construction, modifiers such as *familiar* or *own* would be impossible to interpret.

To summarize, the fact that the postverbal NP is analyzed as a maximal noun phrase argues against the move to treat the verb-plus-*POSS-way* phrase as a complex predicate formed by a lexical rule. One could admittedly take the tack of looking such cases in the eye and calling them lexical items, or perhaps "functional words'" (in the sense of Ackerman & Webelhuth, to appear, or Mohanan 1990; cf. also Zwicky 1990), but this move would in effect equate "lexical item" with what we are here calling "construction." That is, a "lexical item" would be any item that must be listed, that is, any *listeme* in the sense of DiSciullo & Williams 1987 (cf. section 1.1). If this is done, it becomes impossible to distinguish the complex predicate analysis from the alternative proposal made by Jackendoff and defended here, that the *way* examples are instantiations of a particular extralexical construction.

9.6   RELATION TO RESULTATIVES

Marantz (1992) points out that the *way* construction bears a certain similarity to the so-called fake object resultatives discussed in chapter 8 and exemplified by (63):

(63) a.   He cried his eyes red.
      b.   He talked himself hoarse.

In both the *way* construction and this construction, the direct object complement is not normally an argument of the verb. In addition, both constructions disfavor unaccusative verbs (cf. section 9.4.2). Marantz does not propose a specific account of fake object cases, but we have seen that they also admit of a constructional analysis (cf. section 8.5).

Specifically, Marantz claims that "the path named by *way* . . . is the person named by the possessor of *way* extended in space (and time)" (p. 185). This proposal allows Marantz to predict the existence of what I am calling external modification, illustrated by examples (61a–c) above, since on Marantz's account, these cases would actually involve normal internal modification. For example (61c), *They made their weary way home,* involves *weary* modifying *way,* and *way* on Marantz's account is claimed to designate the "movers." Therefore "the *way*" = "the movers" are weary.

It is possible to reinterpret Marantz's claim somewhat to make it stronger. Fake object cases have been argued to preferentially involve reflexives or inalienably possessed terms, specifically body part terms (Jackendoff 1990a). It might be claimed that *way,* while in fact coding the path (and not the mover), designates an inalienably possessed entity. Thus, wherever a person's *way* is, the person must travel. Support for this idea might be drawn from a certain finding in Guaraní (Velásquez-Castillo 1993). Guaraní has a special noun incorporation construction that, with few exceptions, only allows the incorporation of inalienably possessed terms, primarily body part terms. Interestingly, one of the few non–body part terms that is allowed is the term *hape,* translated as "way." (Other non–body part terms include the terms for "talk," "house," and "clothes.") This fact supports the idea that *way* is interpretable as an inalienably possessed path.

What exactly does it mean to say that a path is inalienably possessed? The interpretation that the path (the way) is created by the mover as the mover travels provides an answer: the path exists only where the mover travels because it is created by the traveler. The path is therefore inalienable.

On this reinterpretation of Marantz's proposal, we can no longer directly account for the fact that external modification is allowed in this construction, a fact that was predicted by Marantz. While a full explanation will have to be postponed, it is possible to reduce this problem to a previously unsolved problem. It seems that *way* can occur with external modification even when appearing in another construction—a construction in which *way* clearly does not designate the mover:

(64)   "Rearmament proceeded on its gentle way." (oup)

It is likely that the use of external modification is motivated by the use of *way* meaning "manner," as in the following:

(65)   He spoke in his amusing/eloquent/interminable way.

While it is important to recognize the relationship between the *way* construction and fake object resultatives, it is nonetheless necessary to posit a distinct, albeit related, construction in the grammar to account for the *way* examples. In particular, the following differences between the two constructions remain; these differences prevent us from claiming that the two are actually the *same* construction. As observed above, the *way* construction is available for use with a wide variety of verbs, whereas resultatives in general, and fake object resultatives in particular, are highly restrictive. For example, the fake object resultative analogs of the examples in (7) and (8) are unacceptable:

(66) a. *He bludgeoned himself crazy.
          (with a fake object interpretation wherein he bludgeoned [people in general] until he went crazy)
      b. *He mauled himself silly.
          (meaning that he mauled [people] until he became silly)
      c. *He snorted and injected himself dead.

Moreover, the resultative construction cannot be used to predict the requisite interpretation of *way* expressions that a path is created—that the speaker must construe there to be difficulty or obstacles to the motion. Finally, Dutch is a language which has fake object resultatives, yet does not have the *way* construction (A. Zaenen, personal communication). Because of these various differences, the *way* expressions cannot be directly assimilated to the resultative construction.

9.7 CONCLUSION

This chapter has argued that it is necessary to posit an extralexical grammatical construction in the grammar to account for *way* expressions, since the sentential semantics is not naturally attributed to any of the lexical items' inherent semantics. In particular, no single lexical item can be plausibly assigned responsibility for the motion interpretation or the other semantic constraints detailed in sections 9.3 and 9.4. It is claimed, therefore, that the *way* construction is directly associated with a certain semantics independently of the lexical items which instantiate it. This goes against the current trend of placing an increased emphasis on lexical—particularly verbal—semantics and trying to predict overt complement configuration exclusively from the lexical semantics of the main verb (cf., e.g., Levin & Rapoport 1988; Bresnan & Kanerva 1989; Pinker 1989; Grimshaw 1990).

The analysis of the *way* construction given here extends Jackendoff's similar proposal insofar as the extralexical constructional analysis was argued to be more appropriate than a complex predicate analysis. In addition, the noun *way* has been argued to contribute to the semantics of the means interpretation, instead of being simply a syntactic flag of the construction. On the basis of this, it was suggested that the *way* construction is a conventionalized amalgam of two constructions: the creation construction and the intransitive motion construction. The *way* construction demonstrates the need to recognize *constructional polysemy,* parallel to the polysemy often posited for lexical items and grammatical morphemes.

It has been argued that we need to allow for certain senses of constructions to be more basic (or prototypical) than others. In particular, the means interpretation of the construction was argued to be more basic than the manner interpretation, in that (1) it is accepted by all speakers, whereas judgments about the manner interpretation vary widely, (2) it accounted for 96% of the cases in the data base analyzed, and (3) it was argued to motivate the syntactic form of the construction. Motivation for the manner interpretation was given by noting a similar pattern of polysemy in certain lexical items.

10 Conclusion

This work has been concerned with explicating the nature of argument structure constructions, the relation between verb meaning and constructional meaning, the phenomenon of partial productivity of constructions, and the relation among constructions. Before summarizing the main findings, it is worth discussing some related proposals.

10.1 OTHER CONSTRUCTIONAL APPROACHES

Elements of the constructional approach that has been suggested here are not without precedent, and there are a few voices in the field who have gone against current trends and have noted the need for constructional meaning (cf., e.g., Bolinger 1968; Zwicky 1987, 1989; Zadrozny & Manaster-Ramer 1993). There is also of course previous work within Construction Grammar (Fillmore 1985b, 1987; Lakoff 1987; Fillmore, Kay & O'Connor 1988; Lambrecht 1987, 1990; Brugman 1988; Kay 1990; Michaelis 1993, 1994; Koenig 1993) and the closely related framework of Cognitive Grammar (Langacker 1987a,b, 1988, 1991; Rice 1987b; Kemmer 1988; Tuggy 1988; Maldonado Soto 1992). Coming from a different perspective, Emonds (1991) argues for a "syntactically based semantics" in which syntactic deep-structures are paired with semantic structures.

Other work has explored various means of accounting for the mutual influence of various lexical items in a sentence. For example, MacWhinney (1989) attempts to capture the effects of co-occurring complements on lexical meaning in what he terms "pushy polysemy"; Pustejovsky (1991b) attempts to avoid rampant verbal polysemy by having nouns play a more central role (cf. also Keenan 1984).

I do not attempt to survey the full array of relevant literature here; instead I will briefly discuss how the current proposal is related to the framework suggested by Jackendoff 1990a, the general framework of Montague Grammar (Montague 1973), and that of Wierzbicka 1988.

10.1.1 Jackendoff 1990a

Jackendoff has touched on several of the ideas presented here in *Semantic Structures*. For different reasons, based primarily on the economy of representation as well as the idea that in many cases, an argument is not intuitively a semantic argument of the main verb, Jackendoff suggests extralexical "correspondence rules" to account for examples in which the verb does not lexically code the semantics expressed at the clausal level. At several points he likens these correspondence rules to "constructional idioms," which are—on the present view—constructions: pairings of syntax and semantics that can impose particular interpretations on expressions containing verbs which do not themselves lexically entail the given interpretations.

Jackendoff's discussion of the resultative and *way* constructions, as described in chapters 8–9 above, is particularly close in many respects to the proposals made here. However, there are several differences in perspective and in focus between the two accounts. Many of Jackendoff's correspondence rules are stated as extralexical "adjunct rules." For example, Jackendoff proposes that the postverbal NPs in the following expressions are actually not arguments but adjuncts:

(1)   Bill pushed the piano into the orchestra pit.
(2)   The critics laughed the show out of town. (1990a:233)

However, they fail traditional tests for adjuncthood. For example, they may be passivized:

(3)   The piano was pushed into the orchestra pit.
(4)   The show was laughed out of town.

They occur directly after the verb, without any intervening material:

(5)   *Joe pushed forcefully the piano into the orchestra pit.

And they cannot be left behind if the VP undergoes substitution:

(6)   *Joe pushed the piano into the room and Bill did so the harp.

More generally, Jackendoff (1990a) considers all such complements that are not intuitively licensed by corresponding arguments of the main verb to be adjuncts. In appealing to an argument/adjunct distinction for these cases, the more traditional distinction is rendered obsolete. But then Jackendoff's claim that some direct objects are adjuncts reduces to the claim that some direct objects are not directly associated with an argument of the verb. This is the claim that has been explicitly made here.

We have argued that some direct objects which by traditional tests *do* correspond to arguments are not licensed directly by the verb: these arguments are directly associated with clause-level constructions.[1] Sometimes the arguments associated with a construction are isomorphic with the participants directly associated with the semantics of particular verbs, but sometimes the arguments associated with the construction are imposed on the semantics directly associated with the predicates. This approach allows us to retain the traditional argument/adjunct distinction between subjects, objects, and some PPs on the one hand, and other PPs such as temporal or spatial modifiers on the other hand.

Another difference between Jackendoff's account and the one presented here is that Jackendoff proposes that his adjunct rules apply to sentences "on the fly" to provide them with an interpretation (1990a:235). In the introduction (1990a:9), he draws an analogy between rules that operate "on the fly" and metonymic expressions as analyzed by Nunberg (1979). Nunberg explicitly makes the point that metonymic processes are general and pragmatic—not conventional and not part of grammar. Constructions, on the other hand, as presented here, are by definition conventionalized pieces of grammatical knowledge.[2]

On the present account, constructions play a more central theoretical role than on Jackendoff's account. For Jackendoff, correspondence rules are only required in exceptional cases, when the correspondence rule itself contributes an argument. For the majority of cases, he supposes that argument structure is determined on the basis of *verbal* semantics in isolation. It has been argued here that verbs are generally associated with frame-semantic knowledge that is integrated with independently existing argument structure constructions.

Moreover, in the theory of Construction Grammar, no strict division is drawn between the lexicon and the more general inventory of constructions. Therefore, while Jackendoff claims that his adjunct rules should be considered to operate outside of the lexicon (p. 235), the constructions suggested here can be viewed as free-standing entities, stored within the lexicon alongside lexical items, idioms, and other constructions that may or may not be partially lexically filled.

Other differences between Jackendoff's approach and the present one stem from differences in focus rather than in theoretical perspective. Jackendoff does not attempt to constrain the application of many of his adjunct rules either by adding specific semantic constraints or by delimiting verb classes as has been done here. In fact, he seems to suggest that whether his adjunct rules may apply must be stipulated lexically. This is implied by his analysis of *hit* vs. *strike,* in which he argues that whether a verb can occur with a directional must be stipulated in its individual lexical entry. He makes this point on the basis of the following:

(7)   a.   Bob hit the ball across the field.
       b.   *Joe struck the ball across the field.

Although I have acknowledged there to be some degree of lexical idiosyncrasy (cf. chapter 5), the majority of cases appear to be predictable once a sufficiently detailed semantic characterization of the construction and associated verb classes has been accomplished (see section 7.4.2 for a semantic account of the difference between *hit* and *strike*).

Another difference in focus between the two accounts is that Jackendoff does not attempt to explicitly relate the various constructions that are proposed. One general criticism of Construction Grammar as it has been practiced is that it has rarely attempted to account for the systematic relation among constructions (but cf. Lakoff 1987 and current work by Fillmore and Kay (1993) for notable exceptions). In the enthusiasm to show just how much of language is necessarily learned as idiosyncratic (even where motivated) bits of grammatical knowledge, attention to overarching principles and generalizations has often been lacking. Chapters 3 and 4 of the present work have attempted to address these issues.

10.1.2 Montague Grammar

Montague and many linguists working within the Montague Grammar tradition have adopted the "rule-to-rule" hypothesis (Bach 1976). This approach involves associating each syntactic rule with a semantic rule which determines the meaning of the syntactic constituent formed. Montague Grammar is essentially a system for pairing surface structures with a representation of the meanings of those surface structures, with no significant level of "deep" or "underlying" structure between the two (cf. Montague 1973; Dowty, Wall & Peters 1981). In this way, Construction Grammar and Montague Grammar are quite similar in approach.

One difference between Construction Grammar and Montague Grammar is that the semantic rules in Montague Grammar are supposed to be determined exclusively by the syntactic mode of combination. One cannot refer to semantic features of items being combined in order to determine which semantic rules apply. It is necessary to posit corresponding syntactic features (such as differences in type) in order to constrain the application of the semantic rule. Alternatively, a semantic filter may serve to rule out expressions generated by the erroneous application of a semantic rule. Whether these mechanisms can take the place of explicitly referring to semantics as part of the "mode of combination" is an empirical issue.

Although Montague Grammar has always allowed for the possibility of rich constructional meaning, the actual practice has been to have rules of composition be defined in terms of simple function application. The quotation from Gazdar et al. (1985) cited in chapter 1 reflects that widespread assumption. To repeat their position: "We assume that there exists a universal mapping from syntactic rules to semantic translations. . . . We claim that the semantic type assigned to any lexical item introduced in a rule. . . . and the syntactic form of the rule itself are sufficient to fully determine . . . the form of the semantic translation rule" (1985:8–9).

There may be a recent trend toward assigning richer meanings and semantic constraints to the rules of combination. Such a direction is suggested, for example, by Dowty (1991). Dowty suggests that an alternative to the idea that the unaccusative/unergative distinction is purely grammatical and lexically determined is an analysis in which this distinction is actually an epiphenomenon arising from the semantic constraints on particular constructions. He states: "Certain grammatical constructions have certain meanings associated with them involving P-Agent or P-Patient properties, hence a given intransitive verb is appropriate in such a construction only if it has the right kind of meaning. The set of grammatical rules/constructions appropriate to one semantic class, versus the set appropriate to the other class, thus isolates two classes of verbs, but via semantic constraints originating in the rules themselves" (1991:608). This analysis presupposes that grammatical constructions can be analyzed as having meanings (1991:609).[3]

10.1.3 Wierzbicka 1988

In her book *The Semantics of Grammar,* Wierzbicka argues for the idea that grammatical morphemes and constructions are directly associated with meanings. She motivates this move by noting the sort of systematic semantic distinctions existing in related constructions that were discussed in chapter 1, noting that "language is an integrated system, where everything 'conspires' to convey meaning—words, grammatical constructions, and illocutionary devices (including intonation)" (1988:1).

In arguing that grammatical constructions are directly associated with meaning, Wierzbicka's approach fits squarely into the approach of construction grammar, generally defined. She covers a breathtaking range of data, including causatives crosslinguistically, the Japanese adversative passive, the English ditransitive and a variety of complement types, and particular cases in Polish and Russian. However, the only construction she discusses which is entirely lexically unfilled, and thus directly parallel to the cases discussed here, is the ditransitive construction.

There are other differences between Wierzbicka's work and the present approach. While I have argued that there are lexical exceptions to the generalizations and that there is a high degree of conventionalization associated with the association of verbs and constructions, Wierzbicka argues that the relationship between syntax and semantics is exceptionless. "In every case . . . the syntactic possibilities are determined by the underlying semantic structures (that is, by the intended meaning). Generally speaking, what is semantically incoherent, is syntactically incongruous. Syntax, so to speak, follows from semantics" (1988:4).

Another difference between Wierzbicka's account and the account proposed here is the kind of semantics assumed. She adopts, following Leibnitz, a reductionist approach to semantics, attempting to account for the full range of semantic knowledge associated with open class (and closed class) lexical items in terms of a small set of atomic semantic elements including *I, you, this, someone, something, time, place, want, don't want, say, think, know, imagine, become,* and *part.* She provisionally includes also *like, two, other, world, good, kind of,* and *feel.* She then proposes that the entire semantics of any lexical item can be captured by paraphrases involving these atomic semantic primitives combined in determinate ways. We have taken rather the opposite approach to semantics, arguing that lexical items are associated with rich frame-semantic or encyclopedic knowledge, and that decomposition into atomic elements is impossible.

Finally, the scope of the two projects only overlaps to a limited extent. Wierzbicka concentrates on exemplifying the existence of constructional meaning in a wide variety of constructions and in a wide variety of languages, whereas the present account has focused almost exclusively on causal constructions in English. On the other hand, I have attempted to detail the way verbs and constructions are related, and to provide some overall picture of the way constructions may be related to one another in a hierarchical system.

10.2 SUMMARY

This work has argued, counter to the current trend, that an entirely lexically based approach to grammar is inadequate, and that lexically unfilled constructions must be recognized to exist independently of the particular lexical items which instantiate them. By recognizing the existence of meaningful constructions, we can avoid the claim that the syntax and semantics of the clause is projected exclusively from the specifications of the main verb. In this way, we avoid the problem of positing implausible verb senses to account for examples such as the following:

(8)   He sneezed the napkin off the table.

(9)   She baked him a cake.

(10)   Dan talked himself blue in the face.

In none of these cases does the verb intuitively require the direct object complement. To account for (8), for example, we would have to say that *sneeze,* a parade example of an intransitive verb, actually has a three-argument sense, 'X CAUSES Y to MOVE Z by sneezing'. To account for (9), we would need to claim that there exists a special sense of *bake* that has three arguments—an agent, a theme, and an intended recipient. This in effect argues that *bake* has a sense which involves something like 'X INTENDS to CAUSE Y to HAVE Z by baking'. To account for (10), we would need to postulate a special sense of *talk,* 'X CAUSES Y to BECOME Z by talking'.

On a constructional approach, aspects of the interpretation involving caused motion, intended transfer, or caused result are understood to be contributed by the respective constructions. That is, skeletal argument structure constructions are capable of contributing arguments. For example, the ditransitive construction is directly associated with agent, patient, and recipient roles. We do not need to stipulate a specific sense of *bake* unique to this construction. Thus the direct objects found in examples (8–10) are licensed not directly as arguments of the verb but by the particular constructions. Several other reasons to prefer a constructional account to a lexical approach have been detailed in chapter 1.

In chapter 2, it was argued that verbs must be associated with rich frame-semantic meanings; at the same time, lexical meaning is acknowledged to be highly conventionalized. In particular, which participants associated with a given verb's frame semantics are inherently *profiled* is determined by the lexical specifications of the verb itself. Constructions are also associated with dynamic scenes, but their semantics are more abstract: constructional meanings can be generally captured by skeletal decompositional structures, e.g., 'X CAUSES Y to RECEIVE Z', 'X ACTS', 'X CAUSES Y to MOVE Z', and so on.

It has been argued that constructions as well as lexical items (which are also, strictly speaking, constructions) are typically associated with a family of closely related senses. For example, the ditransitive construction illustrates *constructional polysemy:* the same form is paired with different but related senses. A remarkably similar pattern of extensions was shown to exist for the caused-motion construction (cf. chapters 2, 7). Since polysemy has been shown in many studies to be a natural and recurring phenomenon of lexical items, the existence of constructional polysemy suggests that research which treats constructions and lexical items as the same basic type of data structure, as is done in Construction Grammar, may well be on the right track.

The central senses of argument structure constructions have been argued to be associated with humanly relevant scenes: someone transferring something to someone, something causing something to move or to change state, someone experiencing something, something undergoing a change of state or location, and so on (cf. the "conceptual archetypes" of Langacker 1991). Evidence that these scenes are semantically privileged comes from certain language acquisition facts observed by Clark (1978), Bowerman (1982), and Slobin (1985). Concrete proposals have been made for how to relate verb and construction, and for constraints on that relation.

Systematic metaphors have been shown to play more than a superfluous role in the semantics of constructions. By explicitly recognizing particular metaphors, we can more accurately capture semantic constraints on constructions as well as the relation among constructions.

In chapter 3, various types of relations among constructions have been discussed, including relations of polysemy, metaphorical extension, subsumption, and instance. Constructions are related in an associative network via asymmetric normal mode inheritance links. These inheritance links specify the relation between the dominated and dominating construction; all nonconflicting information is inherited by the dominated construction from the dominating construction. Because the links themselves are treated as objects, they are able to capture the specific nature of the relations among constructions. In addition, the links themselves can be related via inheritance hierarchies of different kinds. Different kinds of links have different type frequencies, and as has been discussed chapter 5, high type frequency is correlated with productivity. Therefore frequently reoccurring links will be used productively, extending familiar patterns in predictable ways.

Chapter 4 presents general arguments for a monostratal approach to the relation between overtly syntactic expression and semantic representations. It has been suggested that the degree of regularity in the relation between semantic role types and overt syntactic expression is sometimes exaggerated, and that many linking generalizations are construction specific. The cross-constructional generalizations that do exist are captured by stating the regularity at a high node in the hierarchy of constructions; subregularities are captured by stating the generalization at an intermediate node. Exceptions are allowed, but only at a cost to the overall system.

The partially productive nature of constructions has been discussed in chapter 5. Adapting insights from Pinker 1989, it has been proposed that constructions are associated with clusters of instances; new or novel cases are acceptable to the degree that they are relevantly similar to existing cases. This usage-based model of grammar was proposed to account for different constructions' varying degrees of productivity (cf. Bybee 1985; Langacker 1987a, 1991). In addition, the idea that there may well be some degree of indirect negative evidence was defended.

While current theories focus almost exclusively on the existence of related constructions, or "alternations," in describing the semantics of particular argument structure expressions, it has been argued that by considering various constructions first on their own terms, instead of immediately focusing on their relations to other constructions, interesting generalizations and subtle semantic constraints emerge. These constraints were detailed in chapters 6–9 for four particular cases: the ditransitive, the caused-motion construction, the resultative, and the *way* construction. Each of these constructions is argued to have independent status, with its own particular semantic constraints, radial category structure, and lexical exceptions, and yet each is shown to be interrelated to other constructions.

Notes

Chapter One

1. Early questions about whether transformations preserve meaning were raised by Kuroda (1965), Partee (1965, 1971), and Bolinger (1968).

2. Wierzbicka suggests that the *to*-plus-infinitive construction is itself associated with the semantic elements "thinking," "wanting," "future time."

3. It is important to bear in mind that both semantic *and* pragmatic aspects of grammatical form are relevant for determining synonymy. Only if two forms have *both* the same semantics and the same pragmatics, they will be disallowed by the Principle of No Synonymy of Grammatical Forms. This principle is impossible to prove conclusively, since one would have to examine all forms in all languages to do so. Further motivation for it is provided in chapter 3.

4. It is not necessary that every syntactic form be uniquely associated with a particular semantics; there are cases of constructional ambiguity, where the same form is paired with distinct meanings. Word order is not part of argument structure constructions, but rather is inherited from other, more general constructions in the language. Thus the statement of the construction, like traditional subcategorization frames, is abstract enough to be instantiated by questions, cleft constructions, and so forth.

5. This definition of course leaves us with a degree of indeterminacy, since there may be several distinct sets of *basis* constructions for a given language. For example, given three patterns, it may turn out that two would be predictable given the third, or that the one would equally well be predictable given the other two. Determining which of the three patterns to consider grammatical constructions is subjective, but the choice will take into account the relative motivations of the various proposed constructions, economy of representation, speaker intuitions about basicness, and so on. It may be possible that speakers actually differ in some cases in the set of grammaticalized constructions they learn, even if their grammar in an extensional sense is the same.

6. "Meaning" is to be construed broadly enough to include contexts of use, as well as traditional notions of semantics. That is, a construction is posited when some aspect of the way it is conventionally used is not strictly predictable. It would alternatively be possible to define constructions as ordered triples of form, meaning, and context as is done by Zadrozny & Manaster-Ramer 1993.

7. With Saussure and Aronoff (1976) among others, I take constructions to be relevantly nonpredictable even if they are partially motivated (cf. section 3.1), as long as they are not strictly or completely predictable.

8. I am using "scene" in the sense of Fillmore (1975, 1977b) to mean an idealization of a "coherent individuatable perception, memory, experience, action, or object" (1977b:84).

9. Carter 1988:171. It should be noted that Carter goes on to argue that natural languages are not strictly compositional.

10. This is not a general criticism of unification grammars, since such grammars are capable in principle of capturing extralexical effects and indirect rules of composition, for

example, by making a distinction between *external* and *internal* semantics. That is, unification grammars can capture these effects by allowing meaning to be contributed by nonlexical nodes. In fact, a current version of Construction Grammar in Fillmore & Kay 1993 adopts a unification system.

11. Although the constraint is not stated in terms of unaccusativity vs. unergativity in chapter 9, the same issue arises: verbs which lexically designate directed motion nevertheless cannot generally occur in the construction:

  (i)   *He descended his way down the stairs.

12. However, extralinguistic knowledge is undoubtedly required as well in order to arrive at a full interpretation of an expression in context; cf. Lakoff 1977, Langacker 1987a.

13. A potential confound of this experiment arises from the possibility of a different degree of bias in examples such as (28a) and (29a). I have not been able to ascertain from the authors whether the degree of bias was controlled for.

14. Unfortunately, only Gleitman 1992 required subjects to use a *set* of syntactic frames to infer verb meaning, and this experiment involved adult speakers identifying already-known verbs, so none of these experiments provide conclusive empirical defense against Pinker's criticism.

15. Fisher et al. (1991) state this idea succinctly: "/touch/ is mapped onto 'touch' because (a) the child can represent scenes observed as 'scenes of touching' and (b) the wave form /touch/ is likely to be heard when touching is happening. That this has to be at least part of the truth about word learning is so obvious as to be agreed upon by all theorists despite their differences in every other regard (see e.g., Locke 1690 and Chomsky 1965—and everybody in between who has commented on the topic). You can't learn a language simply by listening to the radio" (1991 : 2).

16. Even in this case it is clear that *some* amount of word learning is a necessary prerequisite to the learning or identification of constructions.

17. Many lexicalists argue that all "structure-preserving" (Emonds 1972) transformations are better handled in the lexicon (e.g., Freidin 1974, Bresnan 1978). A transformation is considered structure-preserving if (1) both the input and output constructions can be generated by rules of the base component, (2) the forms share a root morphology, and (3) the cooccurrence restrictions of one are predictable from the other. See Wasow 1977 for a more conservative view of the role of lexical rules.

Several proposals involving lexical rules that have been developed recently have questioned many of the traditional distinctions and assumptions usually associated with lexical rule accounts. See, for example, the contributions to Alsina (1994) for discussion. I do not attempt here to compare and contrast these accounts with the current proposal.

18. The question arises as to why *the floor* can't be interpreted metonymically to stand for "a place on the floor" in (32b). But the answer to that question is not a question about ditransitives specifically, since *the floor* in *Joe cleared the floor for Sam* cannot refer to a behind-sized piece of floor either.

Example (32a), *Joe cleared Sam a place on the floor,* is acceptable on the interpretation that a place was *created* for Sam by the clearing (cf. *I wiped him a place on the floor,* which is not acceptable since the place is not created by wiping). It may be that the creation sense of clearing is aided by using the indefinite *a place.* When the definite article is used, the creation interpretation is not as likely, since it is all too clear that the floor and the place on the floor (assuming the metonymy were available) already exist prior to the clearing.

Chapter Two

1. Ali Yazdani (personal communication) points out that it would make more sense, therefore, to call the upper inside of the mouth the *ceiling* of the mouth, as opposed to the *roof* of the mouth, as one does in Persian.

2. This view of verb meaning is parallel to Higginbotham's (1989) notion of the *elucidation* of verb meaning. Higginbotham cites Hale and Keyser's (1985) definition of *cut:* " 'cut' is a V that applies truly to situations *e* involving a patient *y* and an agent *x* who, by means of some instrument *z*, effects in *e* a linear separation in the material integrity of *y*" (Higginbotham, p. 467). (A revised version appeared as Hale & Keyser 1987.) Elucidations are like frame-semantic meanings insofar as they are an attempt to capture the entirety of the meaning associated with a verb. That is, like frame-semantic representations, elucidations do not assume a strict division between dictionary and encyclopedic knowledge. Higginbotham states, "I doubt that a criterial demarcation of lexical and wordly knowledge is necessary, or even desirable, to pose the problems of knowledge and its acquisition that linguistic theory hopes to answer" (p. 470).

3. Recent research within the GB framework has claimed that only the structure and not the content of the theta role array is relevant for syntax (Burzio 1986; Zubizarreta 1987; Rappaport & Levin 1988; Belletti & Rizzi 1988). See Zaenen & Goldberg 1993 for a review of one such approach, that of Grimshaw 1990.

4. The claim that metaphorical extensions are based on the central sense is only intended to entail that *verbs* which must be metaphorically understood represent extensions from the basic sense. As pointed out by Maarten Lemmens (personal communication), not all metaphorical instances of the construction are based on the central sense. For example, (i) involves the metaphorical transfer of a kiss, yet is based on a noncentral sense:

(i)   Mary promised Steve a kiss.

5. Eve Clark (personal communication) states that *give* is also learned early and used relatively frequently.

6. The distinction corresponds roughly to Dowty's (1986) distinction between "individual thematic role" (participant role) and "thematic role type" (argument role). However, for Dowty, thematic role types are determined by intersecting the semantic entailments of all corresponding arguments of a set of predicates; that is, a thematic role type, such as agent or patient, is defined as follows:

> Given a set S of pairs $\langle P, i_P \rangle$, where P is an n-place predicate and i_P is the index of one of its arguments, a *thematic role type* is determined by the intersection of all individual thematic roles determined by S.

Therefore, if there are *m* predicates in a language, then S can be chosen in

$$\sum_{j=1}^{m} \binom{m}{j} = K$$

ways. The total number of role types would only be upperbounded by multiplying K by n, where n is the arity of the predicate with the highest arity. That is to say, there are more than a handful of thematic role types on this view. Dowty suggests that only certain role types such as agent and patient are interesting for linguistics in that they have systematic grammatical consequences; he leaves the determination of which role types are linguistically relevant as an empirical issue. (Cf. Dowty 1991 for a rather different suggestion—the Proto-Agent and Proto-Patient roles he there suggests are not determined by intersecting the set or a particular subset of all entailments of transitive predicates, which might well yield the null set. Cf. discussion in chapter 4.) On the present account, the linguistically relevant "role types" are the roles associated with constructions.

7. Stative verbs must be dealt with differently. I do not attempt to adequately discern their basic meaning here.

8. In some dialects this sentence is in fact acceptable. Such dialects would involve a different lexical entry for *rob*.

9. The verb *rob* normally also implies that the victim is present:

(i) ?Joe robbed her in Hawaii while she was in Chicago on business.

But contexts do occur in which *rob* can be used upon realization of the crime, even if the victim had not been present:

(ii) She walked in the door and realized she'd been robbed!

10. This fact was observed by Jean-Pierre Koenig and Laura Michaelis (personal communication).

11. OBJ$_2$ is the grammatical relation filled by the second NP in ditransitive expressions.

12. The circumstances under which a profiled participant need not be overtly expressed are discussed in section 2.4.5.

13. This parameter may be varied in languages which make extensive use of the applicative construction.

14. Metaphorical extensions such as *She gave him a kick* are discussed in chapter 3.

15. This sketchy analysis of reflexives is inspired by work by Perlmutter (1989) and Alsina (1993). For discussion of broader uses of the reflexive morpheme, see Maldonado Soto (1992) and Kemmer (1988).

16. There are often certain general default inferences about what was eaten (a meal) or drunk (alcohol), but the more specific identity of the referent is unknown or immaterial.

17. It is clear that context does play some role. See Rice 1988 for an interesting discussion.

18. An analogous question is raised by lexical rule accounts. It can be stated: Can a verb from any class be transformed into a verb of any other class?

19. Pinker (1989) similarly notes that Talmy's original example (i) cannot refer to an event in which someone carries a tub of water containing a bottle into a cave (cf. also Carter 1988).

(i) The bottle floated into the cave.

20. Talmy's (1985a) distinction between "means" and "manner" conflation patterns is often misinterpreted. Talmy uses these terms to distinguish: verbs which primarily designate an action performed by an agent, e.g., *push* (the "means" conflation pattern), from those that primarily designate an action of the theme, e.g., *roll* (the "manner" conflation pattern). However, most conflation patterns involving "manner" verbs imply that the particular manner *is* the means of motion. For example, consider (i):

(i) The bottle rolled down the hill.

This entails not only that "the bottle moved down the hill *while* rolling" but also that "the bottle moved down the hill *by* rolling." Therefore I am treating both of these cases as "means" cases and reserving "manner" for verbs which do not encode means.

21. Gruber (1967) and more recently Landau and Gleitman (1985) suggest that in fact *look* is a verb of motion. This analysis relies on a metaphorical interpretation of *look*, whereby the glance of the person looking "travels toward" the thing looked at; there are several difficulties with this analysis (Goldberg 1988); but regardless of the claim made for *look*, it is clear that *aim* is not [+contact], so we need to extend the analysis beyond [+motion, +contact] verbs in any case.

22. Many aspects of these conditions are similar to Matsumoto's (1991) claims as to which types of verbal predicates can be combined to create a complex motion predicate in Japanese. He argues that the complex verbal form is treated as a single word (with respect to argument structure) and discusses the constraints on combinations of verbs as constraints on possible lexicalization patterns (cf. Talmy 1985a). He proposes the following generalization:

An event is semantically conflated with another event in one verb only when
1. the two events share at least one participant and
2. either
   (a) it is the activity or [resulting] state whose duration is coextensive with the duration of the other event, or
   (b) it is the cause of the other event, or the means with which the other event is caused.

However, the constraints on the Japanese complex predicate construction are actually somewhat less stringent than the constraints on English. For example, Matsumoto cites (i) below, noting that this example can be uttered felicitously when arriving at a coffee house after opening a window back at the office.

(8)   Boku wa mado o   akete kita.
       I TOP   window ACC   open came
       'I came after opening a window.'

The relation between the opening and the arriving is only one of temporal proximity; this type of relation between e_v and e_c is not possible in English. The fact that Japanese is freer in its "conflation patterns" is likely attributable at least in this type of case to the fact that there are two verbs in the Japanese construction.

Chapter Three

1. It should be pointed out that the relation in form must be interpreted as *representing* in some way the relation in meaning. We would not expect a relation in meaning to be motivated by just any relation in form. As Clark, Gelman and Lane (1985) point out, compound formation is a familiar way (in English and many other languages) to encode subordinate-level categories.

2. Some authors have conflated the notions of inheritance and object-oriented design since the two often co-occur in particular implementations. However, the ideas are conceptually distinct.

3. Polysemy links and instance links are relevantly like Langacker's "extension" and "elaboration" links, respectively (see Langacker 1987a, 1988).

4. An anonymous reviewer points out that *drive* appears with other resultative phrases as well, if the resultative is encoded as a PP:

(i)   Chris drove Pat to suicide/to drink.

These cases indicate that "crazy" is perhaps too narrow, and that the meaning of the result-goal argument is better described as "to a state where normal mental processing is not possible."

5. At the same time, a subpart link does not necessarily entail the existence of an instance link; there exist subpart links between certain constructions which do not involve one construction being an instance of another construction. For example, the caused-motion construction and the intransitive motion construction are related by a subpart relation since the intransitive motion construction is a proper subpart of the caused-motion construction; yet the caused-motion construction is not an instance of the intransitive motion construction.

6. L. Levin, Mitamura, and Mahmoud (1988), as discussed by Rappaport Hovav and Levin (1991), assume a metaphorical relation between the two constructions, although they apparently do not provide explicit arguments for such an analysis.

7. At least one verb, *render*, seems to further require that the resultative phrase be an adjective:

> (i)   a. The catastrophe rendered her helpless/ineffectual/impotent.
>   b. *The catastrophe rendered her out of commission.

This at first led me to consider that a further division should be drawn between adjectival and prepositional resultative phrases; however, *render* appears to be an isolated case and is better accounted for by a lexical stipulation linking the result-goal argument to an AP. In particular, the lexical entry for *render* will link the result-goal argument to an AP.

8. Gruber and Jackendoff do not actually refer to a metaphor. Instead they propose that the domains of ownership and physical transfer share an abstract shema (see also Langacker 1987a for a similar view). An approach involving metaphors is preferred here because of the asymmetric nature of the relation between change of ownership and physical transfer. While we find many words that are "basically" associated with physical motion being used in the domain of change of ownership, we do not find instances of the reverse. Moreover, physical transfer is more directly understood than the more abstract domain of transfer of ownership in that the former is directly perceivable. See Lakoff and Johnson 1980 for further arguments against an abstractionist account of similar phenomena.

9. They do differ in whether the recipient argument role of the construction is profiled or not—whether it is expressed by a direct grammatical function; however they are semantically the same in designating 'X CAUSES Y to RECEIVE Z'.

10. A definition of focus that is adequate for our purposes is found in Halliday (1967): "Information focus is one kind of emphasis, that whereby the speaker marks out a part . . . of a message block as that which he wishes to be interpreted as informative. What is focal is 'new' information; not in the sense that it cannot have been previously mentioned, although it is often the case that it has not been, but in the sense that the speaker presents it as not being recoverable from the preceding discourse" (p. 204). Cf. Lambrecht (1994) for a recent thorough discussion of this notion.

11. Although (37) and (38) are acceptable, they seem to be slightly less preferred than their ditransitive counterparts, indicating that the caused-motion construction generally tends to prefer the goal argument to be focused and the transferred object nonfocused.

12. This is of course not to say that *every* potential instance of a systematic metaphor can appear in a particular construction, but only that the particular instances which are otherwise conventional will occur in the construction, as long as the semantic and pragmatic conditions are satisfied.

13. It is true that (40a–42a) and (44a–45a) are more marked than (35). This might be because (35), repeated here as (i), is sufficiently like the non-metaphorical case in (ii) to be less noticeably an instance of metaphorical extension.

(i) #She gave a brand-new house to him.

(ii) She gave a nickel to him.

Chapter Four

1. However, as Van Valin (1992) points out, what counts as "reasonably productive morphology" is not adequately defined, since Baker suggests that the output of an incorporation transformation can be regular, irregular, or even suppletive morphologically.

2. Since the world is necessarily filtered through our cognitive and perceptual apparatus, what we really have access to are *construals* of situations. This idea is by no means new, and extends at least as far back as Hobbes.

3. Baker does not seem to recognize the difference in interpretation between these two sentences since, citing Fodor 1970, he explicitly rejects analyzing *kill* from an underlying *cause to die* because the two predicates are not synonymous. It seems clear that the same semantic difference is involved in the Chicheŵa examples (Van Valin 1992).

4. This is a weaker claim than Dryer made in 1983, when he argued that a single rule showing a difference between two kinds of arguments is sufficient evidence that the two are distinct grammatical relations (Dryer 1983:139).

5. DO and IO are not replaced by PO and SO; rather, Dryer leaves it an empirical issue which of these grammatical relations exist in a particular language.

6. Dryer (1986) also cites the following example:

(i) a. John baked a pie for Mary.
    b. John baked Mary a pie.

He again assumes that the two forms must share a level of representation because of their close semantic relationship; in particular, he assumes that one must be derived from the other. The question thus arises which form is derived from which. In order to decide, Dryer relies implicitly on the UAH. "The beneficiary nominal in [b] . . . behaves like a final term in its case-marking and position. *Since its semantic properties would suggest that it is an initial non-term,* it must have advanced to become a term" [italics added]. If, because of semantic distinctions, we no longer accept as given the idea that there necessarily *is* an earlier stage of derivation, Dryer's account is immediately undermined.

7. There are of course overgeneralizations, but these conform to the general semantic constraints of the construction.

8. This is done, of course, at the cost of complicating the relationship between underlying and surface forms.

9. Some traditional grammars identify the first object of ditransitive expressions (20) with the OBL complement of expressions such as (21), classifying both as syntactically Indirect Objects. This move, which is motivated primarily by considerations of semantic similarity, is cogently argued against by Faltz (1978) and Hudson (1992).

10. If we assume that this lexical rule is a nongenerative redundancy rule, then both entries must be stored in the lexicon, and the arrow is better represented as pointing in both directions:

$$cook_0 \: \langle agt \: theme \rangle \leftrightarrow cook_1 \: \langle agt \: \theta_{dependent} \: theme \rangle$$

11. In order to account for the English ditransitive in a similar way, they would need to suppose that the ditransitive did not necessarily add an argument, but could also alter an existing argument to be a "dependent" argument and thus able to receive the intrinsic classification of $[-r]$.

12. In a subsequent paper, Ackerman (1992) assigns the theme argument the intrinsic classification of $[-o]$. This account moves away from previous ones in that it treats intrinsic classifications as being assigned according to entailments of the predicates (à la Dowty 1991) and adds the notion *markedness* to the mapping theory, but it still leaves us with a range of "intrinsic classifications" consisting of $[-r]$, $[+o]$, and $[-o]$ for the theme argument, depending on which construction is to be expressed.

13. Chicheŵa seems to allow a very similar locative alternation as English at least for some verbs (examples from Sam Mchombo, personal communication):

(i)   a-na-pachira mchenga m'ngolo
       SM-PST-load 3-sand   in-9-cart
       'He loaded the sand in the cart.'

(ii)   a-na-pachira ngolo ndi mchenga
       SM-PST-load 9-cart with sand
       'He loaded the cart with sand.'

14. In the case of *undergo, sustain,* and *tolerate,* the participant realized as subject is causally affected (a Proto-Patient property), while the participant realized as the object causes the affect (a Proto-Agent property). In the case of *receive, inherit,* the participant expressed as subject is stationary relative to the movement of another participant (a Proto-Patient property), while the participant realized as the object moves (a Proto-Agent property).

15. See Hopper & Thompson 1980 for some debate about whether the prototypical patient is inanimate.

16. Of course one must be careful to avoid classifying certain sentences as syntactically transitive solely because they encode a semantically prototypical transitive event. Otherwise this crosslinguistic generalization could turn out to be vacuous.

Chapter Five

1. The epigraph was taken from a squib by Arnold Zwicky in which he lists twenty properties that are systematically associated with manner of speaking verbs. Unfortunately the explanation alluded to in the quote is not forthcoming, as the quote is the last paragraph in the squib.

2. I thank Jess Gropen for bringing this possibility to my attention.

3. Examples of the latter situation would seem to include the English middle construction (e.g., *This book reads easily*) and the *way* construction (cf. chapter 9). However, these particular cases are perhaps not convincing as cases which require recourse to Pinker's subclasses, since they seem to be fully productive once general semantic constraints on the constructions are identified.

4. Steven Pinker (personal communication) has suggested that differences in judgment are only unexpected if they do not also hold for the input forms. For example, we should not expect (i.a) to be better than (i.b):

(i)   a. ?She blasted him a cannonball.
       b. ?She blasted a cannonball to him.

And in fact, both examples are fairly odd.

However, the question still remains as to why the input form is not fully acceptable, since other verbs of ballistic motion are acceptable in that form. How is it that the child learns that *blast* is not completely felicitous in the input argument structure? Moreover, this line of reasoning will not account for the difference in judgments between (30) and (31), since in both cases the input forms are completely acceptable:

(ii)   Sally designed a sculpture for him.

(iii)   Sally created a sculpture for him.

5. More needs to be said about how exactly these cases would be worked out on a constructional account. I do not attempt a full explanation here.

6. It is not clear whether this information is stored indefinitely, since Gropen et al.'s finding of a tendency toward conservatism was only demonstrated in a single experimental encounter. It would be interesting to see if this tendency toward conservativism were lessened by allowing an interval of a number of days to intervene between the acquisition of a novel word and the subsequent production of that word.

7. I have not attempted to apply a formal similarity metric so the relative closeness of the circles is not claimed to be accurate in detail.

8. The frequency of *way* examples is increased dramatically—to one example in every 2,500 words—in one particular subtext of the Lund corpus (not included in the above statistic) taken from various sports commentaries. The difference in frequency can be attributed to the semantic constraints on the construction: forceful or deliberate motion despite obstacles is particularly likely to be described in competitive sport contexts.

Chapter Six

1. Subjects which metonymically stand for volitional beings are also acceptable:

(i)   His company promised him a raise.

(ii)   The orchestra played us the symphony.

2. I would like to thank Dirk Geeraerts (personal communication) and Alan Schwartz (personal communication) for indicating that this metaphor could be stated in terms of transfer.

3. Many theories capture this constraint by postulating a beneficiary role for the first object position of expressions that are paraphrasable with a benefactive *for*-phrase.

4. Examples (23) and (24) happen to be based on metaphors. What is relevant here is that successful (metaphorical) transfer is implied: (23) implies that Chris has a headache, and (24) implies that Chris got a kick.

Chapter Seven

1. See Napoli 1992 for some discussion of the possible existence of a resultative construction in Italian.

2. Gawron is less clear about how the PP and verb are to be syntactically joined. At one point he specifies that co-predicators are subcategorized for by the main verb (1986:328). Thus, the PP phrase would be syntactically subcategorized for although not an argument of the verb. Later (1986:368) Gawron suggests that instead the PP might be added by lexical rule or as an adjunct.

3. See also Gawron 1986, Carrier & Randall 1992, and Hoekstra & Mulder 1990 for arguments against treating the PP or adjective of the resultative as an adjunct.

4. Hoekstra (1992) does not address cases of resultatives which are applied to the direct objects of transitive verbs put to their normal transitive use. In an earlier paper (Hoekstra 1988) he discusses the possibility of a "stripping" rule that removes all arguments normally associated with a predicate before the resultative attaches. If this rule is assumed, his account could extend to cover cases of transitive resultatives.

5. Rappaport Hovav and Levin are often cited as positing a Direct Object Requirement (DOR) constraint on resultatives. However, recognizing that the NP of a small clause is not a direct object, they revise this constraint in the second half of their 1991 paper, to allow for resultatives based on (unergative) intransitive verbs. Their final formulation is that resultatives are restricted to apply to an argument which is governed by the verb.

6. By choosing the term "e-position," Hoekstra seems to be alluding to a Davidsonian event variable.

7. This claim is factually incorrect. Although the majority of resultatives clearly involve atelic verbs, telic predicates also allow resultatives:

(i)   a. She frightened him off his rocker.
      b. He broke the walnuts into the bowl.
      c. Sam closed the door shut.

8. This view of coercion is somewhat different than that proposed by Sag and Pollard. Sag and Pollard propose a rule of coercion that operates on particular lexical items but does not make reference to any licensing construction. The view presented here is preferred since it constrains the potentially all-powerful process by requiring that constructions coerce lexical items into having systematically related meanings.

9. In fact, Carter (1988) has proposed such a construction.

10. An anonymous reviewer of an earlier draft of this chapter pointed out that *lure* does seem to allow *willingly* to be predicated of the theme argument:

(i)   He$_i$ was lured into the room willingly$_i$.

This can be interpreted as "He allowed himself to be lured into the room." I have no account of why this case is different than the others.

11. It might be objected that *strike* can be used when the impacting entity, and not the impacted entity is affected:

(i)   The mosquito struck the window. (example from Paul Kay)
(ii) The car struck a brick wall.

However, in these cases the argument whose location is in question—i.e., the theme—is the subject, not the direct object. The following example is ruled out by the Unique Path Constraint (cf. section 3.4.1), since the car and the wall would have to be interpreted as moving along two distinct paths.

(iii) *The car struck a brick wall into pieces.

12. The distinction has also been cast, equivalently for our purposes, as one betweeen "ballistic" and "controlled" causation (Shibatani 1973).

13. Their analysis differs somewhat from the one presented here in that they assume that the difference in interpretation stems from a different sense of the verb, not from a difference in construction.

14. Pinker includes *wad* in this class, but I would not define *wad* as "to force a mass into

a container." I also do not find *wad* acceptable in either construction: **He wadded the hole with tissues/*He wadded tissues into the hole.*

15. Pinker includes a sixth class: "Mass is caused to move in a widespread or nondirected distribution: *bestrew, scatter, sow, strew.*" (1989:126). However, I don't find any of these examples acceptable in the causative variant: **Joe scattered the field with seeds.*

16. Many of the verbs in this class have a different sense in which the theme role is necessarily volitional. This sense can occur in the intransitive motion construction:

(i)   The crowd crammed/jammed/packed/crowded into the auditorium.

17. There is a potential problem with this characterization of the *spray* class since it seems to violate the Principle of Correspondence posited in section 2.4.2. According to that principle, profiled participant roles must fuse with profiled argument roles (except in the case of a third profiled participant role, which is allowed to fuse with a nonprofiled argument role). Recall that profiled argument roles are those roles expressed by direct grammatical functions. Now, in the examples in (127), we find one of the two profiled roles, liquid or target, fusing with a nonprofiled argument role (expressed as an oblique). This problem can be solved in one of two ways. On the one hand, one might restrain the Principle of Correspondence so that it can be overridden by the Principle of Semantic Coherence, which states that any two roles that are fused must be semantically compatible. That is, as the participant roles of *spray* are fused with the argument roles of the caused-motion construction, for example, the Principle of Semantic Coherence will prevent both the liquid and the target role from fusing with the profiled causer role of the construction: neither role can be construed as an instance of the causer role. Therefore, in order to fuse successfully, the Principle of Correspondence must be overridden to allow one of the *liquid* or target roles to fuse with the oblique role.

Alternatively, one could resort to positing two distinct senses for each verb of the *spray*-class: one sense would have all participant roles profiled, the other would profile only the liquid and target roles. The two senses would be related since they share the same background frame, differing only in the number of profiled roles.

Chapter Eight

1. The interpretation of volitionality is not a hard-and-fast constraint, however; speakers find *Those rolls overbake easily* to be acceptable. (I thank Annie Zaenen for bringing this example to my attention.)

2. I thank Jane Espenson for suggesting this example.

3. This follows from the fact that in Role and Reference Grammar, the framework developed by Van Valin, English passive is stated as an operation on the undergoer.

4. To see how the account could be translated into a semantics-changing lexical rule type of account, see Goldberg 1991a.

5. Note that if the verb's patient-type participant role is profiled, then it *must* be fused with the patient argument role of the construction; if it is not profiled, then the construction does not rule out the possibility that it is left unexpressed, and that the patient role is contributed by the construction.

6. It is possible that example (16) in the text, repeated below, is also a novel extension based on the idea that "they" became fat to the point of being nonfunctional.

(i)   Whose whole life is to eat, and drink . . . and laugh themselves fat. (OED: Trapp, Comm. and Epist. and Rev. (1947))

7. Napoli (1992) has independently suggested a similar constraint that is argued to hold even more strongly in Italian. Because I received her manuscript in the final stages of writing this chapter, I have not attempted to compare and contrast our accounts.

Chapter Nine

1. I am grateful to Patrick Hanks for compiling these examples, and to Annie Zaenen for forwarding them to me.

2. The distinction between means and manner that is used here is slightly different than Talmy's (1985a) distinction between "means" and "manner" conflation patterns. Talmy used these terms only to distinguish verbs which primarily designate an action performed by an agent (e.g., *push*) from those that primarily designate an action of the theme (e.g., *roll*) in sentences such as the following:

  (i)   a. Joe pushed the barrel down the hill.
  (ii)   Joe rolled the barrel down the hill.

However, both of the verbs in the above examples would be classified as designating the means of motion for our purposes. In particular, *roll* as well as *push* must designate the means of motion, and cannot designate a contingent coextensive manner. Notice that (ii) could not be used felicitously in the circumstance in which the barrel is being rolled between Joe's hands as Joe walked down the hill (cf. Pinker 1989 and Croft 1991 for further examples of this point). That is, the rolling must crucially be the means of motion (as well as designating a particular manner).

3. There existed in the OED other uses of *way* as a direct object previous to this date; for example, *After the enterment the kyng tok his way* (1338), *childe fiet ne dar guo his way vor fie guos fiet blaufi* (1340). However, these cases are instances of a different construction: the path phrase is not obligatory as it is today, and the interpretation is quite different: the predicates in these examples meant, roughly, either "to go away" or "to take one's leave." Notice that the verbs *take* and *go* are no longer acceptable in the *way* construction at all: **He went/ took his way to the beach.*

4. I thank Michael Israel for this observation.

5. Notice (45) is not excluded simply because it involves use of the verb *jump,* since the following variant is acceptable:

  (i)   She jumped her way over the ditches.

This is because, as Jackendoff notes, *jump* is acceptable in this construction just in case it is interpreted iteratively (1990a:224).

6. It seems that for some speakers, myself included, this constraint is strengthened in the case of human movers to a constraint that the motion must be volitional (although the motion may terminate at an unintended location). For example, *She tripped her way down the stairs* is not acceptable for some speakers.

7. Jackendoff provides the following example, however:

  (i)   The barrel rolled its way up the alley. (1990a:212)

I myself find this example unacceptable, and I suspect that Jackendoff may have had a personification interpretation in mind because he further includes the following:

  (ii) The barrel rolled its ponderous way up the alley.

He paraphrases (ii) as (iii):

(iii)   The barrel, ponderous (as an elephant), went up the alley rolling. (1990a:217)

8. I thank Charles Fillmore for this observation.

Chapter Ten

1. An anonymous reviewer points out that showing that a phrase is not an adjunct is not sufficient to show that it is an argument, since there exist cases of nonsemantic complements, for instance, in raising constructions. However, it is clear that the complements in the present cases are semantically constrained (cf. chapters 7–8). Therefore these cases are unlike raising cases, and are in fact arguments.

2. Jackendoff (personal communication) has said that his intention was not to imply that the adjunct rule was not a conventional part of grammar by saying that the adjunct rules were interpreted "on the fly." In a more recent paper (Jackendoff 1992), he argues in fact that Nunberg's examples are also not purely pragmatic, in the sense of being outside of the grammar (cf. also Na 1986, and in fact Nunberg himself in Nunberg & Zaenen 1992 for arguments that certain metonymies are language specific). Therefore, although it was not made clear in the original text, Jackendoff's actual view is that the correspondence rules are conventionalized pieces of grammar.

3. A concrete proposal along these lines, albeit in a different framework, has been made by Legendre, Miyata, and Smolensky (1991). They propose that each test frame for unaccusativity may be associated directly with its own semantic requirements; they further propose that *in addition* each verb lexically encodes a binary syntactic feature which designates whether the verb is unaccusative. They suggest that grammaticality is determined by allowing for the interaction of semantic constraints of the various constructions with the syntactic marker of unaccusativity.

Bibliography

Abbreviations

BLS n   *Proceedings of the* nth *Annual Meeting of the Berkeley Linguistics Society,* University of California, Berkeley

CLS n   *Papers from the* nth *Annual Regional Meeting of the Chicago Linguistic Society,* University of Chicago

Ackerman, Farrell. 1990. Locative Inversion vs. Locative Alternation. *Proceedings of the 9th West Coast Conference on Formal Linguistics,* 1–14.

Ackerman, Farrell. 1992. Complex Predicates and Morphological Relatedness: Locative Alternation in Hungarian. In Ivan A. Sag and Anna Szabolcsi, eds., *Lexical Matters.* CSLI Lecture Notes no. 24, 55–84. Stanford, Cal.: Center for the Study of Language and Information, Stanford University.

Ackerman, Farrell, and Gert Webelhuth. To appear. *Wordhood and Syntax: The Theory of Complex Predicates.* Stanford, Cal.: Center for the Study of Language and Information, Stanford University.

Aissen, Judith. 1983. Indirect Object Advancement in Tzotzil. In D. Perlmutter, ed., *Studies in Relational Grammar,* vol. 1, 272–302. Chicago: University of Chicago Press.

Alsina, Alex. 1992. On the Argument Structure of Causatives. *Linguistic Inquiry* 23(4): 517–555.

Alsina, Alex. 1993. *Predicate Composition: A Theory of Syntactic Function Alternations.* Ph.D. diss., Stanford University.

Alsina, Alex, ed. 1994. *Complex Predicates.* Stanford, Cal.: Center for the Study of Language and Information, Stanford University.

Alsina, Alex, and Sam Mchombo. 1990. The Syntax of Applicatives in Chicheŵa: Problems for a Theta Theoretic Asymmetry. *Natural Language and Linguistic Theory* 8(4): 493–506.

Anderson, John R. 1984. Spreading Activation. In J. R. Anderson and S. M. Kosslyn, eds., *Tutorials in Learning and Memory.* San Francisco: W. H. Freeman.

Anderson, Stephen R. 1971. On the Role of Deep Structure in Semantic Interpretation. *Foundations of Language* 6: 197–219.

Aronoff, Mark. 1976. *Word Formation in Generative Grammar.* Linguistic Inquiry Monograph 1. Cambridge, Mass.: MIT Press.

Aske, Jon. 1989. Motion Predicates in English and Spanish: A Closer Look. *BLS* 15, 1–14.

Austin, John L. 1940. The Meaning of a Word. Reprinted in *Philosophical Papers.* Oxford: Oxford University Press, 1961.

Bach, Emmon. 1976. An Extension of Classical Transformational Grammar. In *Problems of Linguistic Metatheory* (proceedings of the 1976 conference). East Lansing: Michigan State University.

Baker, C. L. 1979. Syntactic Theory and the Projection Problem. *Linguistic Inquiry* 10: 533–581.

Baker, Mark C. 1987. Incorporation and the Nature of Linguistic Representation. Paper presented at conference, The Role of Theory in Language Description, Ocho Rios, Jamaica.

Baker, Mark C. 1988. *Incorporation: A Theory of Grammatical Function Changing.* Chicago: University of Chicago Press.

Bartlett, Frederick. 1932. *Remembering.* Cambridge: Cambridge University Press.

Bates, Elizabeth, and Brian MacWhinney. 1987. Competition, Variation and Language Learning. In Brian MacWhinney, ed., *Mechanisms of Language Acquisition,* 157–193. Hillsdale, N.J.: Lawrence Erlbaum Associates.

Belletti, Adriana, and Luigi Rizzi. 1988. Psych-verbs and Theta Theory. *Natural Language and Linguistic Theory* 6:291–352.

Bloom, L. 1970. *Language development: Form and Function in Emerging Grammars.* Cambridge, Mass.: MIT Press.

Bloom, L., and M. Lahey. 1978. Language Development and Language Disorders. New York: Wiley.

Bloom, L., P. Miller, and L. Hood. 1975. Variation and Reduction as Aspects of Competence in Language Development. In A. Pick, ed., *Minnesota Symposia on Child Development,* vol. 9, 3–55. Minneapolis: University of Minnesota Press.

Bobrow, D. G., and T. Winograd. 1977. An Overview of KRL, a Knowledge Representation Language. *Cognitive Science* 1(1):3–46.

Bobrow, D. G., and B. Webber. 1980. Knowledge Representation of Syntactic/Semantic Processing. In *Proceedings of the First National Conference on Artificial Intelligence,* 316–323. San Mateo, Cal.: Morgan Kaufmann.

Bolinger, Dwight L. 1968. Entailment and the Meaning of Structures. *Glossa* 2: 119–127.

Bolinger, Dwight L. 1971. *The Phrasal Verb in English.* Cambridge, Mass.: Harvard University Press.

Borkin, Ann. 1974. *Problems in Form and Function.* Ph.D. diss., *University of Michigan.* Published, Norwood, N.J.: Ablex Publishing, 1984.

Bowerman, Melissa. 1973. *Early Syntactic Development: A Cross-linguistic Study with Special Reference to Finnish.* Cambridge: Cambridge University Press.

Bowerman, Melissa. 1982. Reorganizational Processes in Lexical and Syntactic Development. In E. Wanner and L. R. Gleitman, eds., *Language Acquisition: The State of the Art,* 319–346. Cambridge: Cambridge University Press.

Bowerman, Melissa. 1988. The 'No Negative Evidence' Problem: How Do Children Avoid Constructing an Overly General Grammar? In J. Hawkins, ed., *Explaining Language Universals,* 73–101. Oxford: B. Blackwell.

Bowerman, Melissa. 1989. Learning a Semantic System: What Role Do Cognitive Predispositions Play? In M. L. Rice and R. L. Schiefelbusch, eds., *The Teachability of Language,* 133–169. Baltimore: P. H. Brookes.

Braine, M. D. S. 1971. On Two Types of Models of the Internalization of Grammars. In D. I. Slobin, ed., *Ontogenesis of Grammar.* New York: Academic Press.

Braine, M. D. S., R. E. Brody, S. M. Fisch, and M. J. Weisberger. 1990. Can Children Use a Verb without Exposure to Its Argument Structure? *Journal of Child Language* 17:313–342.

Bresnan, Joan. 1969. On Instrumental Adverbs and the Concept of Deep Structure. MIT Quarterly Progress Report 92. MIT, Cambridge, Mass.

Bresnan, Joan. 1978. A Realistic Transformational Grammar. In M. Halle, J. Bresnan, and G. A. Miller, eds. *Linguistic Theory and Psychological Reality,* 1–59. Cambridge, Mass.: MIT Press.

Bresnan, Joan. 1982. *The Mental Representation of Grammatical Relations.* Cambridge, Mass.: MIT Press.

Bresnan, Joan. 1990. Levels of Representation in Locative Inversion: A comparison of English and Chicheŵa. Invited address presented at the 13th GLOW Colloquium at Cambridge University. Revised and duplicated Stanford University, Stanford, Cal.

Bresnan, Joan, and Jonni Kanerva. 1989. Locative Inversion in Chicheŵa. *Linguistic Inquiry* 20:1–50.

Bresnan, Joan, and Lioba Moshi. 1990. Object Asymmetries in Comparative Bantu Syntax. *Linguistic Inquiry* 21(2):147–185.

Bresnan, Joan, and Annie Zaenen. 1990. Deep Unaccusativity in LFG. In K. Dziwirek et al., eds., *Grammatical Relations: A Cross-Theoretical Perspective,* 45–57. Stanford, Cal.: Center for the Study of Language and Information, Stanford University.

Brown, Roger. 1957. Linguistic Determinism and Parts of Speech. *Journal of Abnormal and Social Psychology* 55:1–5.

Brown, Roger. 1973. *A First Language: The Early Stages.* Cambridge, Mass.: Harvard University Press.

Brown, Roger, and Camille Hanlon. 1970. Derivational Complexity and Order of Acquisition in Child Speech. In J. R. Hays, ed., *Cognition and the Development of Language,* 11–53. New York: Wiley.

Brugman, Claudia M. 1981. *The Story of 'Over': Polysemy, Semantics, and the Structure of the Lexicon.* Master's thesis, University of California, Berkeley. Published, New York: Garland, 1988.

Brugman, Claudia M. 1988. *The Syntax and Semantics of 'have' and Its Complements.* Ph.D. diss., University of California, Berkeley.

Burzio, Luigi. 1986. *Italian Syntax: A Government and Binding Approach.* Dordrecht: Reidel.

Bybee, Joan. 1985. *Morphology: A Study of the Relation between Meaning and Form.* Amsterdam: Benjamins.

Carlson, Greg N., and Michael K. Tanenhaus. 1988. Thematic Roles and Language Comprehension. In W. Wilkins, ed., *Syntax and Semantics 21: Thematic Relations,* 263–288. New York: Academic Press.

Carrier, Jill, and Janet H. Randall. 1992. The Argument Structure and Syntactic Structure of Resultatives. *Linguistic Inquiry* 23:173–234.

Carroll, J., P. Davies, and B. Richman. 1971. *Word Frequency Book.* New York: Houghton Mifflin.

Carter, Richard. 1988. Compositionality and Polysemy. In B. Levin and C. Tenny, eds., *On Linking: Papers by Richard Carter,* 167–204. MIT Lexicon Project Working Paper no. 25. Department of Linguistics and Philosophy, MIT, Cambridge, Mass.

Cattell, Ray. 1984. *Syntax and Semantics 17: Composite Predicates in English*. New York: Academic Press.

Channon, Robert. 1980. On Place Advancements in Russian and English. In C. V. Chvany and R. D. Brecht, eds., *Morphosyntax in Slavic,* 114–138. Columbus, Ohio: Slavica Publishers.

Chomsky, Noam. 1957. *Syntactic Structures*. The Hague: Mouton.

Chomsky, Noam. 1965. *Aspects of the Theory of Syntax*. Cambridge, Mass.: MIT Press.

Chomsky, Noam. 1970. Remarks on Nominalization. In R. Jacobs and P. Rosenbaum, eds., *Readings in English Transformational Grammar*. Waltham, Mass.: Ginn.

Chomsky, Noam. 1981. *Lectures on Government and Binding*. Dordrecht: Foris.

Chomsky, Noam. 1986. *Knowledge of Language*. New York: Praeger.

Chomsky, Noam. 1992. *A Minimalist Program for Linguistic Theory*. MIT Occasional Papers in Linguistics 1. Cambridge, Mass.: Dept. of Linguistics and Philosophy, MIT.

Clark, Eve V. 1978. Discovering What Words Can Do. In *Papers from the Parasession on the Lexicon, CLS 14,* 34–57.

Clark, Eve V. 1987. The Principle of Contrast: A Constraint on Language Acquisition. In B. MacWhinney, ed., *Mechanisms of Language Acquisition,* 1–33. Hillsdale, N.J.: Lawrence Erlbaum Associates.

Clark, Eve V., and Herb H. Clark. 1979. When Nouns Surface as Verbs. *Language* 55: 767–811.

Clark, Eve V., Susan A. Gelman, and Nancy Lane. 1985. Compound Nouns and Category Structure in Young Children. *Child Development* 56: 84–91.

Comrie, Bernard. 1984. Subject and Object Control: Syntax, Semantics, and Pragmatics. *BLS* 10, 450–464.

Croft, William. 1991. *Syntactic Categories and Grammatical Relations*. Chicago: University of Chicago Press.

Davis, Anthony. 1993. Linking, Inheritance and Semantic Structures. Presentation at the Center for the Study of Language and Information, Stanford University.

DeLancy, Scott. 1991. Event Construal and Case Role Assignment. *BLS* 17, 338–353.

Dinsmore, John. 1979. *Pragmatics, Formal Theory and the Analysis of Presupposition*. Ph.D. diss., University of California, San Diego.

DiSciullo, Anna-Maria, and Edwin Williams. 1987. *On the Definition of Word*. Cambridge, Mass.: MIT Press.

Dixon, R. M. W. 1972. *The Dyirbal Language of North Queensland*. Cambridge: Cambridge University Press.

Dowty, David. 1972. *Studies in the Logic of Verb Aspect and Time Reference in English*. Studies in Linguistics, Department of Linguistics, University of Texas, Austin.

Dowty, David. 1979. *Word Meaning and Montague Grammar*. Dordrecht: Reidel.

Dowty, David. 1986. Thematic Roles and Semantics. *BLS* 12, 340–354.

Dowty, David. 1991. Thematic Proto-Roles and Argument Selection. *Language* 67(3): 547–619.

Dowty, David, Robert Wall, and Stanley Peters. 1981. *Introduction to Montague Semantics*. Dordrecht: Reidel.

Dryer, Matthew. 1983. Indirect Objects in Kinyarwanda Revisited. In D. M. Perlmutter, ed., *Studies in Relational Grammar,* vol. 1, 129–140. Chicago: University of Chicago Press.

Dryer, Matthew. 1986. Primary Objects, Secondary Objects, and Antidative. *Language* 62(4):808–845.

Emanatian, Michele. 1990. The Chagga Consecutive Construction. In J. Hutchison and V. Manfredi, eds., *Current Approaches to African Linguistics,* vol. 7, 193–207. Dordrecht: Foris Publications.

Emonds, Joseph. 1972. Evidence That Indirect Object Movement Is a Structure-Preserving Rule. *Foundations of Language* 8:546–561.

Emonds, Joseph. 1991. Subcategorization and Syntax-Based Theta-Role Assignment. *Natural Language and Linguistic Theory* 9(3):369–429.

England, Nora. 1983. A Grammar of Mam: A Mayan Language. Austin: University of Texas Press.

Erteschik-Shir, Nomi. 1979. Discourse Constraints on Dative Movement. In T. Givón, ed., *Syntax and Semantics 12: Discourse and Syntax,* 441–467. New York: Academic Press.

Fahlman, S. 1979. *NETL: A System for Representing and Using Real-World Knowledge.* Cambridge, Mass.: MIT Press.

Faltz, Leonard. 1978. On Indirect Objects in Universal Syntax. *CLS* 14, 76–87.

Farrell, Patrick. 1991. *Thematic Relations, Relational Networks and Multistratal Representations.* Ph.D. diss., University of California, San Diego.

Fauconnier, Gilles. 1985. *Mental Spaces.* Cambridge, Mass.: MIT Press.

Filip, Hana. 1993. *Aspect, Situation Type and Nominal Reference.* Ph.D. diss., University of California, Berkeley.

Fillmore, Charles J. 1968. The Case for Case. In E. Bach and R. T. Harms, eds., *Universals in Linguistic Theory,* 1–88. New York: Holt, Rinehart and Winston.

Fillmore, Charles J. 1970. The grammar of Hitting and Breaking. In R. Jacobs and P. Rosenbaum, eds., *Readings in English Transformational Grammar,* 120–133. Waltham, Mass.: Ginn.

Fillmore, Charles J. 1971. Some Problems for Case Grammar. In R. O'Brian, ed., *Report on the Twenty-Second Annual Round Table Meeting on Languages and Linguistics.* Washington: Georgetown University Press.

Fillmore, Charles J. 1975. An Alternative to Checklist Theories of Meaning. *BLS* 1, 123–131.

Fillmore, Charles J. 1976. Frame Semantics and the Nature of Language. In S. Harnad, H. Steklis, and J. Lancaster, eds., *Origins and Evolutions of Language and Speech.* New York: New York Academy of Sciences.

Fillmore, Charles J. 1977a. The Case for Case Reopened. In P. Cole, ed., *Syntax and Semantics 8: Grammatical Relations,* 59–81. New York: Academic Press.

Fillmore, Charles J. 1977b. Topics in Lexical Semantics. In R. Cole, ed., *Current Issues in Linguistic Theory,* 76–138. Bloomington: Indiana University Press.

Fillmore, Charles J. 1982. Frame Semantics. In Linguistic Society of Korea, ed., *Linguistics in the Morning Calm,* 111–138. Seoul: Hanshin.

Fillmore, Charles J. 1985a. Frames and the Semantics of Understanding. *Quaderni di Semantica* 6(2):222–53.

Fillmore, Charles J. 1985b. Syntactic Intrusions and the Notion of Grammatical Construction. *BLS* 11, 73–86.

Fillmore, Charles J. 1986. Pragmatically Controlled Zero Anaphora. *BLS* 12, 95–107.

Fillmore, Charles J. 1987. Lectures held at the Stanford Summer Linguistics Institute, Stanford University.

Fillmore, Charles J. 1988. The Mechanisms of "Construction Grammar." *BLS* 14, 35–55.

Fillmore, Charles J. 1990. Construction Grammar. Course reader for Linguistics 120A, University of California, Berkeley.

Fillmore, Charles J., and Paul Kay. 1993. Construction Grammar. Unpublished manuscript, University of California, Berkeley.

Fillmore, Charles J., Paul Kay, and Catherine O'Connor. 1988. Regularity and Idiomaticity in Grammatical Constructions: The Case of *Let Alone*. *Language* 64: 501–538.

Fisher, Cynthia, Geoffrey Hall, Susan Rakowitz, and Lila Gleitman. 1991. When it Is Better to Receive than to Give: Syntactic and Conceptual Constraints on Vocabulary Growth. IRCS Report 91–41. Philadelphia: University of Pennsylvania.

Flickinger, Daniel, Carl Pollard, and Thomas Wasow. 1985. Structure-Sharing in Lexical Representation. In *Proceedings of the 23rd Annual Meeting of the Association for Computational Linguistics,* 262–267. Chicago: Association for Computational Linguistics.

Fodor, Jerold A. 1970. Three Reasons for Not Deriving *Kill* from *Cause to Die*. *Linguistic Inquiry* 1:429–438.

Fodor, Jerold A., Janet D. Fodor, and Merrill F. Garrett. 1975. The Psychological Unreality of Semantic Representations. *Linguistic Inquiry* 6:515–531.

Fodor, Jerold A., Merrill F. Garrett, E. C. T. Walker, and C. H. Parkes. 1980. Against Definitions. *Cognition* 8:263–267.

Foley, William A., and Robert Van Valin, Jr. 1984. *Functional Syntax and Universal Grammar*. Cambridge Studies in Linguistics 38. Cambridge: Cambridge University Press.

Frege, Gottlob. 1979. Begriffsschrift, a formula language, modeled upon that of arithmetic, for pure thought. In J. van Heijenoort, ed. (1970), *Frege and Gödel: Two Fundamental Texts in Mathematical Logic*. Cambridge, Mass.: Harvard University Press.

Freidin, Robert. 1974. Transformations and Interpretive Semantics. In R. Shuy and N. Bailey, eds., *Towards Tomorrow's Linguistics,* 12–22. Washington, D.C.: Georgetown University Press.

Gawron, Jean Mark. 1983. *Lexical Representations and the Semantics of Complementation*. Ph.D. diss., University of California, Berkeley. Published, New York: Garland, 1988.

Gawron, Jean Mark. 1985. A Parsimonious Semantics for Prepositions and CAUSE. *CLS* 21, Part 2, *Papers from the Parasession on Causatives and Agentivity,* 32–47.

Gawron, Jean Mark. 1986. Situations and Prepositions. *Linguistics and Philosophy* 9(4):427–476.

Gazdar, Gerald, Ewan Klein, Geoffrey Pullum, and Ivan Sag. 1985. *Generalized Phrase Structure Grammar*. Cambridge, Mass.: Harvard University Press.

Gelman, Susan A., Sharon A. Wilcox, and Eve V. Clark. 1989. Conceptual and Lexical Hierarchies in Young Children. *Cognitive Development* 4(4):309–326.

Gergely, Gyorgy, and Thomas G. Bever. 1986. Related Intuitions and the Mental Representation of Causative Verbs in Adults and Children. *Cognition* 23:211–277.

Gibbs, Ray. 1990. Psycholinguistic Studies on the Conceptual Basis of Idiomaticity. *Cognitive Linguistics,* vol. 1. New York: Mouton de Gruyer.

Gibson, J. J. 1950. *The Perception of the Visual World.* Boston: Houghton Mifflin.

Givón, Talmy. 1979a. *On Understanding Grammar.* New York: Academic Press.

Givón, Talmy. 1979b. From Discourse to Syntax: Grammar as a Processing Strategy. In T. Givón, ed., *Syntax and Semantics 12: Discourse and Syntax.* New York: Academic Press.

Givón, Talmy. 1985. Function, Structure, and Language Acquisition, In D. I. Slobin. ed. *The Crosslinguistic Study of Language Acquisition,* vol. 2, 1005–1028. Hillsdale, N.J.: Lawrence Erlbaum Associates.

Gleitman, Lila. 1992. Presentation at the 15th International Congress of Linguists, Quebec City, Canada.

Goldberg, Adele E. 1988. Semantic Roles of Statives in RRG. Unpublished manuscript, University of California, San Diego.

Goldberg, Adele E. 1991a. A Semantic Account of Resultatives. *Linguistic Analysis* 21(1–2):66–96.

Goldberg, Adele E. 1991b. It Can't Go Up the Chimney Down: Paths and the English Resultative. *BLS* 17, 368–378.

Goldberg, Adele E. 1992a. The Inherent Semantics of Argument Structure: The Case of The English Ditransitive Construction. *Cognitive Linguistics* 3(1):37–74.

Goldberg, Adele E. 1992b. *Argument Structure Constructions.* Ph.D. diss., University of California, Berkeley.

Goldsmith, John. 1980. Meaning and Mechanism in Language. In S. Kuno, ed. *Harvard Studies in Syntax and Semantics,* vol. 3. Cambridge, Mass.: Harvard University Press.

Goldsmith, John. 1993. Harmonic Phonology. In J. Goldsmith, ed., *The Last Phonological Rule: Reflections on Constraints and Derivations,* 21–60. Chicago: University of Chicago Press.

Gordon, David, and George Lakoff. 1971. Conversational Postulates. *CLS* 7, 63–84.

Green, Georgia. 1972. Some Observations on the Syntax and Semantics of Instrumental Verbs. *CLS* 8, 83–97.

Green, Georgia. 1973. A Syntactic Syncretism in English and French. In B. Kachru et al., eds., *Issues in Linguistics,* 257–278. Urbana: University of Illinois Press.

Green, Georgia. 1974. *Semantics and Syntactic Regularity.* Bloomington: Indiana University Press.

Greenfield, P. M., and J. Smith. 1976. *The Structure of Communication in Early Language Development.* New York: Academic Press.

Grégoire, A. 1937. *L'apprentissage du langage,* vol. 1. Paris: Droz.

Grimshaw, Jane. 1979. Complement Selection and the Lexicon. *Linguistic Inquiry* 10(2):279–326.

Grimshaw, Jane. 1990. *Argument Structure.* Cambridge, Mass.: MIT Press.

Gropen, Jess, Steven Pinker, Michelle Hollander, Richard Goldberg, and Ronald Wilson. 1989. The Learnability and Acquisition of the Dative Alternation in English. *Language* 65(2):203–257.

Gropen, Jess, Steven Pinker, Michelle Hollander, and Richard Goldberg. 1991. Affectedness and Direct Objects: The Role of Lexical Semantics in the Acquisition
of Verb Argument Structure. *Cognition* 41:153–195.

Gruber, Jeffrey S. 1965. *Studies in Lexical Relations.* Ph.D. diss., MIT.

Gruber, Jeffrey S. 1967. *Look* and *See. Language* 43:937–947.

Guerssel, M., K. Hale, M. Laughren, B. Levin, and J. White Eagle. 1985. A Crosslinguistic Study of Transitive Alternations. *CLS* 21, Part 2, *Papers from the
Parasession on Causatives and Agentivity,* 48–63.

Guillaume, Paul. 1927. The Development of Formal Elements in the Child's Speech.
Reprinted in C. Ferguson and D. Slobin, eds., *Studies in Child Language
Development.* New York: Holt, Rinehart and Winston.

Haiman, John. 1978. A Study in Polysemy. *Studies in Language* 2(1):1–34.

Haiman, John. 1985. *Natural Syntax: Iconicity and Erosion.* Cambridge: Cambridge
University Press.

Hale, Ken, and Samuel Keyser. 1985. The View from the Middle. Unpublished manuscript, MIT, Cambridge, Mass.

Hale, Ken, and Samuel Keyser. 1987. A View from the Middle. Lexicon Project Working Papers 10. Center for Cognitive Science, MIT.

Halliday, Michael A. K. 1967. Notes on Transitivity and Theme in English. *Journal of
Linguistics* 3:199–244.

Higginbotham, James. 1989. Elucidations of Meaning. *Linguistics and Philosophy* 12:
465–517.

Hoekstra, Teun. 1988. Small Clause Results. *Lingua* 74:101–139.

Hoekstra, Teun. 1992. Aspect and Theta Theory. In I. M. Roca, ed., *Thematic Structure:
Its Role in Grammar.* Berlin: Mouton de Gruyter.

Hoekstra, Teun, and René Mulder. 1990. Unergatives as Copular Verbs: Locational and
Existential Production. *The Linguistic Review* 7:1–79.

Hopper, P. J., and S. A. Thompson. 1980. Transitivity in Grammar and Discourse. *Language* 56: 251–299.

Hudson, Richard. 1984. *Word Grammar.* Oxford: Basil Blackwell.

Hudson, Richard. 1992. So-called 'Double Objects' and Grammatical Relations. *Language* 68(2):251–276.

Jackendoff, Ray. 1969. An Interpretive Theory of Negation. *Foundations of Language*
5(2):218–241.

Jackendoff, Ray. 1972. *Semantic Interpretation in Generative Grammar.* Cambridge,
Mass.: MIT Press.

Jackendoff, Ray. 1975. Morphological and Semantic Regularities in the Lexicon. *Language* 51(3):639–671.

Jackendoff, Ray. 1983. *Semantics and Cognition.* Cambridge, Mass.: MIT Press.

Jackendoff, Ray. 1987. *Consciousness and the Computational Mind.* Cambridge,
Mass.: MIT Press.

Jackendoff, Ray. 1990a. *Semantic Structures.* Cambridge, Mass.: MIT Press.

Jackendoff, Ray. 1990b. On Larson's Treatment of the Double Object Construction.
Linguistic Inquiry 21(3):427–455.

Jackendoff, Ray. 1992. Mme. Tussaud Meets the Binding Theory. *Natural Language
and Linguistic Theory* 10:1–31.

Jakobson, Roman. 1938. *Russian and Slavic Grammar.* Reprinted in L. R. Waugh and

M. Halle, eds., *Janua Linguarum,* Series Major 106. New York: Mouton, 1984.

Jespersen, Otto. 1949. *A Modern English Grammar on Historical Principles.* Part 6, *Syntax.* Copenhagen: Munksgaard.

Jurafsky, Daniel. 1992. An On-line Computational Model of Human Sentence Interpretation: A Theory of the Representation and Use of Linguistic Knowledge. Ph.D. diss., University of California, Berkeley, and Report No. UCB/CSD 92/676, Computer Science Division, University of California, Berkeley.

Kapur, Shyam. 1993. How Much of What? Is This What Underlies Parameter Setting? In *Proceedings of the 25th Annual Stanford Language Research Forum,* 50–59. Stanford, Cal.: Center for the Study of Language and Information, Stanford University.

Katz, Jerry J., and Paul Postal. 1964. *An Integrated Theory of Linguistic Descriptions.* Cambridge, Mass.: MIT Press.

Katz, N., E. Baker, and J. McNamara. 1974. What's in a Name? *Child Development* 45: 469–473.

Kay, Martin. 1984. Functional Unification Grammar: A Formalism for Machine Translation. In *Proceedings of the International Conference on Computational Linguistics,* 75–78.

Kay, Paul. 1990. Even. *Linguistics and Philosophy* 13(1):59–112.

Keenan, Edward L. 1972. On Semantically Based Grammar. *Linguistic Inquiry* 4(3): 413–462.

Keenan, Edward L. 1976. Towards a Universal Definition of "Subject." In C. N. Li, ed., *Subject and Topic,* 303–334. New York: Academic Press.

Keenan, Edward L. 1984. Semantic Correlates of the Ergative/Absolutive Distinction. *Linguistics* 22:197–223.

Kemmer, Suzanne. 1988. *The Middle Voice.* Ph.D. diss., Stanford University.

Kiparsky, Paul. 1987. Morphology and Grammatical Relations. Unpublished manuscript, Stanford University.

Kirsner, Robert S. 1985. Iconicity and Grammatical Meaning. In J. Haiman, ed., *Iconicity in Syntax,* 249–270. Amsterdam: John Benjamins.

Koenig, Jean-Pierre. 1993. Linking Constructions vs. Linking Rules: Evidence from French. *BLS* 19, 217–231.

Kuroda, Sige-Yuki. 1965. *Generative Grammatical Studies in the Japanese Language.* Ph.D. diss., MIT.

Lakoff, George. 1965. *On the Nature of Syntactic Irregularity.* Ph.D. diss., Indiana University. Published as *Irregularity in Syntax.* New York: Holt, Rinehart and Winston, 1970.

Lakoff, George. 1968. Instrumental Adverbs and the Concept of Deep Structure. *Foundations of Language* 4:4–29.

Lakoff, George. 1970a. Adverbs and Opacity: A Reply to Stalnaker. Unpublished manuscript, University of California, Berkeley.

Lakoff, George. 1970b. Global Rules. *Language* 46:627–639.

Lakoff, George. 1971. On Generative Semantics. In D. D. Steinberg and L. A. Jakobovits, eds., *Semantics: An Interdisciplinary Reader in Philosophy, Linguistics and Psychology,* 232–296. London: Cambridge University Press.

Lakoff, George. 1972. Linguistics and Natural Logic. In D. Davidson and G. Harman, eds., *Semantics of Natural Language,* 545–665. Dordrecht: Reidel.

Lakoff, George. 1976. Towards Generative Semantics. In J. D. McCawley, ed., *Syntax and Semantics 7: Notes from the Linguistic Underground,* 43–62. New York: Academic Press. First circulated in 1963.

Lakoff, George. 1977. Linguistic Gestalts. *CLS* 13, 225–235.

Lakoff, George. 1984. *There*-Constructions: A Case Study in Grammatical Construction Theory and Prototype Theory. Cognitive Science Technical Report 18, University of California, Berkeley. Revised version published as "Case Study" in Lakoff 1987.

Lakoff, George. 1987. *Women, Fire, and Dangerous Things: What Categories Reveal about the Mind.* Chicago: University of Chicago Press.

Lakoff, George. 1993. The Contemporary Theory of Metaphor. In A. Ortony, ed., *Metaphor and Thought,* 2d ed. Cambridge: Cambridge University Press.

Lakoff, George, and Mark Johnson. 1980. *Metaphors We Live By.* Chicago: University of Chicago Press.

Lakoff, George, and John Robert Ross. 1976. Is Deep Structure Necessary? In J. D. McCawley, ed., *Syntax and Semantics 7: Notes from the Linguistic Underground,* 159–164. New York: Academic Press. First circulated in 1967.

Lakoff, Robin. 1968. *Abstract Syntax and Latin Complementation.* Cambridge, Mass.: MIT Press.

Lambrecht, Knud. 1987. Sentence Focus, Information Structure, and the Thetic–Categorical Distinction. *BLS* 13, 366–382.

Lambrecht, Knud. 1990. 'What me worry?' Mad Magazine Sentences Revisited. *BLS* 16, 215–228.

Lambrecht, Knud. 1994. *Information Structure and Sentence Form: A Theory of Topic, Focus, and the Mental Representation of Discourse Referents.* Cambridge Studies in Linguistics. Cambridge: Cambridge University Press.

Landau, Barbara, and Lila R. Gleitman. 1985. *Language and Experience: Evidence from the Blind Child.* Cambridge, Mass.: Harvard University Press.

Langacker, Ronald W. 1969. Pronominalization and the Chain of Command. In D. Reibel and S. Schane, eds., *Modern Studies in English: Readings in Transformational Grammar.* Englewood Cliffs, N.J.: Prentice-Hall.

Langacker, Ronald W. 1985. Observations and Speculations on Subjectivity. In J. Haiman, ed., *Iconicity in Syntax,* 109–150. Amsterdam: John Benjamins.

Langacker, Ronald W. 1987a. *Foundations of Cognitive Grammar.* Vol. 1: *Theoretical Prerequisites.* Stanford, Cal.: Stanford University Press.

Langacker, Ronald W. 1987b. Grammatical Ramifications of the Setting/Participant Distinction. *BLS* 13, 383–394.

Langacker, Ronald W. 1988. An Overview of Cognitive Grammar. In B. Rudzka-Ostyn, ed. *Topics in Cognitive Linguistics,* 127–161. Philadelphia: John Benjamins.

Langacker, Ronald W. 1991. *Foundations of Cognitive Grammar.* Vol. 2: *Descriptive Application.* Stanford, Cal.: Stanford University Press.

Laughren, Mary. 1988. Toward a Lexical Representation of Warlpiri Verbs. In W. Wilkins, ed., *Syntax and Semantics 21: Thematic Relations,* 215–242. New York: Academic Press.

Legendre, Géraldine, Yoshiro Miyata, and Paul Smolensky. 1990. Harmonic Grammar—A Formal Multi-level Connectionist Theory of Linguistic Well-formedness: Theoretical Foundations. Technical report no. 90-5, Institute of Cognitive Science, University of Colorado at Boulder.

Legendre, Géraldine, Yoshiro Miyata, and Paul Smolensky. 1991. Unifying Syntactic and Semantic Approaches to Unaccusativity: A Connectionist Approach. *BLS* 17, 156–167.

Le Roux, C. 1988. On the Interface of Morphology and Syntax: Evidence from Verb Particle Combinations in Afrikaans. *Stellenbosch Papers in Linguistics* 18. University of Stellenbosch, South Africa.

Levin, Beth. 1985. Lexical Semantics in Review: An Introduction. In B. Levin, ed., *Lexical Semantics in Review*. Lexicon Project Working Papers 1. Cambridge, Mass.: MIT Center for Cognitive Science.

Levin, Beth. 1993. *English Verb Classes and Alternations*. Chicago: University of Chicago Press.

Levin, Beth, and T. Rapoport. 1988. Lexical Subordination. *CLS* 24, Part 1, 275–289.

Levin, Beth, and Malka Rappaport. 1986. The Formation of Adjectival Passives. *Linguistic Inquiry* 17:623–661.

Levin, Beth, and Malka Rappaport Hovav. 1990a. The Lexical Semantics of Verbs of Motion: The Perspective from Unaccusativity. In I. M. Roca, ed., *Thematic Structure: Its Role in Grammar,* 247–269. Berlin: Mouton de Gruyter.

Levin, Beth, and Malka Rappaport Hovav. 1990b. Wiping the Slate Clean: A Lexical Semantic Exploration. *Cognition* 41:123–155.

Levin, Beth, and Malka Rappaport Hovav. 1992. Classifying Single Argument Verbs. Unpublished manuscript, Northwestern University and Bar Ilan University.

Levin, Lori. 1987. Toward a Linking Theory of Relation Changing Rules in LFG. Report CSLI-87-115. Stanford, Cal.: Center for the Study of Language and Information, Stanford University.

Levin, Lori, T. Mitamura, and A. T. Mahmoud. 1988. Lexical Incorporation and Resultative Secondary Predicates. Presentation at the LSA Annual Meeting, New Orleans.

Lieber, Rochelle. 1988. Phrasal Compounds in English and the Morphology–Syntax Interface. *CLS* 24, 202–222.

Lindner, Susan. 1981. *A Lexico-Semantic Analysis of Verb-Particle Constructions with* Up *and* Out. Ph.D. diss., University of California, San Diego.

Locke, John. 1690. An Essay Concerning Human Understanding. Published 1964. Cleveland: Meridian Books.

McCawley, James D. 1968a. The Role of Semantics in a Grammar. In E. Bach and R. T. Harms, eds., *Universals in Linguistic Theory,* 124–169. New York: Holt, Rinehart and Winston.

McCawley, James D. 1968b. Lexical Insertion in a Transformational Grammar without Deep Structure. *CLS* 4, 71–80.

McCawley, James D. 1973. Syntactic and Logical Arguments for Semantic Structures. In O. Farjimura, ed., *Three Dimensions in Linguistic Theory,* 259–376. Tokyo: TEC Corporation.

McCawley, James D., ed. 1976. *Syntax and Semantics 7: Notes from the Linguistic Underground.* New York: Academic Press.

McCawley, James D. 1978. Conversational Implicature and the Lexicon. In P. Cole, ed., *Syntax and Semantics 9: Pragmatics,* 245–259. New York: Academic Press.

McCawley, James D. 1986. The Focus and Scope of *Only.* University of Chicago Working Papers in Linguistics. Linguistics Department, University of Chicago.

McClelland, J. L., D. E. Rumelhart, and G. E. Hinton. 1986. The Appeal of Parallel Distributed Processing. In Rumelhart and McClelland, eds., 1986, vol. 1: *Foundations. Parallel Distributed Processing,* 3–44.

MacWhinney, Brian. 1978. *The Acquisition of Morphophonology.* Monographs of the Society for Research in Child Development, vol. 43. Chicago: University of Chicago Press.

MacWhinney, Brian. 1989. Competition and Lexical Categorization. In R. Corrigan, F. Eckman, and M. Noonan, eds., *Current Issues in Linguistic Theory. Vol. 61: Linguistic Categorization.* Amsterdam Studies in the Theory and History of Linguistic Science, series 4. Amsterdam: John Benjamins.

MacWhinney, Brian. 1991. Connectionism as a Framework for Language Acquisition Theory. In J. Miller, *Research on Child Language Disorders,* 73–104. Austin, Texas: Pro-Ed.

Makkai, Adam. 1972. *Idiom Structure in English.* The Hague: Mouton.

Maldonado Soto, Ricardo. 1992. Middle Voice: The Case of Spanish *Se.* Ph.D. diss., University of California, San Diego.

Marantz, Alec P. 1984. *On the Nature of Grammatical Relations.* Cambridge, Mass.: MIT Press.

Marantz, Alec P. 1992. The *Way*-construction and the Semantics of Direct Arguments in English: A Reply to Jackendoff. In T. Stowell and E. Wehrli, eds., *Syntax and Semantics 26: Syntax and the Lexicon,* 179–188. New York: Academic Press.

Maratsos, M., R. Gudeman, P. Gerard-Ngo, and G. DeHart. 1987. A Study in Novel Word Learning: The Productivity of the Causative. In B. MacWhinney, ed., *Mechanisms of Language Acquisition.* Hillsdale, N.J.: Lawrence Erlbaum Associates.

Matsumoto, Yo. 1991. Some Constraints on the Semantic Structures of Verbs: Evidence from Japanese Motion Predicates. Unpublished manuscript, Stanford University.

Mchombo, Sam. 1978. *A Critical Appraisal of the Place of Derivational Morphology in Transformational Grammar.* Ph.D. diss., School of Oriental and African Studies, University of London.

Mchombo, Sam. 1992. The Stative in Chichewa and the Relevance of Thematic Information. Presentation at the Cognitive Science Colloquium, University of California, Berkeley.

Meyer, D. E., and R. W. Schvaneveldt. 1971. Facilitation in Recognizing Pairs of Words: Evidence of a Dependence between Retrieval Operations. *Journal of Experimental Psychology* 90:227–234.

Michaelis, Laura. 1993. *Toward a Grammar of Aspect: The Case of the English Perfect Construction.* Ph.D. diss., University of California, Berkeley.

Michaelis, Laura. 1994. A Case of Constructional Polysemy in Latin. *Studies in Language* 18:23–48.

Minsky, Marvin. 1975. A Framework for Representing Knowledge. In P. H. Winston, ed., *The Psychology of Computer Vision.* New York: McGraw-Hill.

Mithun, Marianne. 1991. Active/Agentive Case Marking and Its Motivation. *Language* 67(3):510–547.

Mohanan, Tara. 1990. *Arguments in Hindi.* Ph.D. diss., Stanford University.

Montague, Richard. 1973. The Proper Treatment of Quantifiers in Ordinary English. In J. Hintikka, J. Moravecsik, and P. Suppes, eds., *Approaches to Natural Language.* Dordrecht: Reidel. Reprinted in R. Thomason, ed., *Formal Philosophy: Selected Papers of Richard Montague.* New Haven: Yale University Press, 1974.

Morolong, Mailillo, and Larry Hyman. 1977. Animacy, Objects, and Clitics in Sesotho. *Studies in African Linguistics* 8:199–217.

Mufwene, Salikoko. 1978. English Manner-of-Speaking Verbs Revisited. *CLS* 14, *Papers from the Parasession on the Lexicon,* 278–289.

Na, Younghee. 1986. The Conventionalization of Semantic Distinctions. *CLS* 22, Part 1, 166–178.

Naigles, Letitia. 1990. Children Use Syntax to Learn Verb Meanings. *Journal of Child Language* 357–374.

Naigles, Letitia, Henry Gleitman, and Lila Gleitman. 1993. Children Acquire Word Meaning Components from Syntactic Evidence. In E. Dromi, ed., *Language and Cognition: A Developmental Perspective,* 104–140. Norwood, N.J.: Ablex Publishing.

Napoli, Donna Jo. 1992. Secondary Resultative Predicates in Italian. *Journal of Linguistics* 28:53–90.

Norvig, Peter, and George Lakoff. 1987. Taking: A Study in Lexical Network Theory. *BLS* 13, 195–206.

Nunberg, Geoffrey. 1979. The Non-Uniqueness of Semantic Solutions: Polysemy. *Linguistics and Philosophy* 3(2):143–184.

Nunberg, Geoffrey, and Annie Zaenen. 1992. Systematic Polysemy in Lexicology and Lexicography. In Hannu Tommola, Krista Varantola, Tarja-Salmi-Tolonen and Jurgen Schopp, eds., *Proceedings of Euralex II.* Tampere, Finland: University of Tampere.

Nunberg, Geoffrey, Thomas Wasow, and Ivan Sag. 1982. Idioms: An Interim Report. In *Proceedings of the Plenary Sessions, 13th International Congress of Linguists,* Tokyo 1982.

Oehrle, Richard T. 1976. *The Grammatical Status of the English Dative Alternation.* Ph.D. diss., MIT.

Oosten, Jeanne van. 1977. Subjects and Agenthood in English. *CLS* 13, 459–471.

Oosten, Jeanne van. 1984. *The Nature of Subjects, Topics and Agents: A Cognitive Explanation.* Ph.D. diss., University of California, Berkeley. Distributed by Indiana University Linguistics Club, Bloomington.

The Compact Edition of the Oxford English Dictionary (OED). 1971. Oxford: Oxford University Press.

Park, T.-Z. 1977. Emerging Language in Korean Children. Unpublished MS., Institute of Psychology, Bern.

Partee, Barbara Hall. 1965. *Subject and Object in Modern English.* Published in J. Hankamer, ed., Outstanding Dissertations in Linguistics Series. New York: Garland, 1979.

Partee, Barbara Hall. 1971. On the Requirement That Transformations Preserve Meaning. In C. J. Fillmore and D. T. Langendoen, eds., *Studies in Linguistic Semantics,* 1–21. New York: Holt, Rinehart and Winston.

Perlmutter, David M. 1978. Impersonal Passives and the Unaccusative Hypothesis. *BLS* 4, 157–189.

Perlmutter, David M. 1989. Multiattachment and the Unaccusative Hypothesis: the Perfect Auxiliary in Italian. *Probus* 1(1):63–119.

Perlmutter, David M., and Paul M. Postal. 1983a. Some Proposed Laws of Basic Clause Structure. In D. M. Perlmutter, eds., *Studies in Relational Grammar,* vol. 1, 81–128. Chicago: University of Chicago Press.

Perlmutter, David M., and Paul M. Postal. 1983b. Toward a Universal Characterization of Passivization. In D. M. Perlmutter, ed., *Studies in Relational Grammar,* vol. 1, 3–29. Chicago: University of Chicago Press.

Pinker, Steven. 1981. Comments on the paper by Wexler. In C. L. Baker and J. J. McCarthy, eds., *The Logical Problem of Language Acquisition,* 53–78. Cambridge, Mass.: MIT Press.

Pinker, Steven. 1984. *Language Learnability and Language Development.* Cambridge, Mass.: Harvard University Press.

Pinker, Steven. 1987. The Bootstrapping Problem in Language Acquisition. In B. MacWhinney, ed., *Mechanisms of Language Acquisition,* 399–441. Hillsdale, N.J.: Lawrence Erlbaum Associates.

Pinker, Steven. 1989. *Learnability and Cognition: The Acquisition of Argument Structure.* Cambridge, Mass.: MIT Press.

Pinker, Steven, and Alan Prince. 1988. On Language and Connectionism: Analysis of a Parallel Distributed Processing Model of Language Acquisition. *Cognition* 28:73–193.

Pinker, Steven, and Alan Prince. 1991. Rules and Associations. *BLS* 17, 230–251.

Pinker, Steven, D. S. Lebeaux, and L. A. Frost. 1987. Productivity and Constraints in the Acquisition of the Passive. *Cognition* 26:195–267.

Pollard, Carl, and Ivan A. Sag. 1987. *Information-Based Syntax and Semantics 1: Fundamentals.* CSLI Lecture Notes Series no. 13. Stanford, Cal.: Center for the Study of Language and Information, Stanford University.

Pollard, Carl, and Ivan A. Sag. 1994. *Head-Driven Phrase Structure Grammar.* Chicago: University of Chicago Press and Stanford, Cal.: Center for the Study of Language and Information.

Postal, Paul M. 1971. On the Surface Verb *Remind.* In C. J. Fillmore and D. T. Langendoen, eds., *Studies in Linguistic Semantics,* 181–270. New York: Holt, Rinehart and Winston. Also in *Linguistic Inquiry* 1:37–120 (1970).

Prince, Alan, and Paul Smolensky. 1991. Lecture Notes on Connectionism and Linguistic Theory. Distributed at the Summer Institute of Linguistics, Santa Cruz, Cal.

Pustejovsky, James. 1991a. The Syntax of Event Structure. *Cognition* 41:47–81.

Pustejovsky, James. 1991b. The Generative Lexicon. *Computational Linguistics* 17(4): 409–441.

Quillian, M. Ross. 1968. Semantic Memory. In M. Minsky, eds., *Semantic Information Processing,* 227–270. Cambridge, Mass.: MIT Press.

Quine, W. V. O. 1960. *Word and Object.* Cambridge, Mass.: MIT Press.

Randall, Janet H. 1983. A Lexical Approach to Causatives. *Journal of Linguistic Research* 2(3):77–105.

Rappaport, Malka, and Beth Levin. 1985. A Study in Lexical Analysis: The Locative Alternation. Unpublished manuscript, Bar Ilan University and Northwestern University.

Rappaport, Malka, and Beth Levin. 1988. What to Do with Theta Roles. In W. Wilkins, eds., *Syntax and Semantics 21 Thematic Roles,* 7–36. New York: Academic Press.

Rappaport Hovav, Malka, and Beth Levin. 1991. Is There Evidence for Deep Unaccusativity in English? An Analysis of Resultative Constructions. Unpublished manuscript, Bar Ilan University and Northwestern University.

Ratcliff, R., and G. McKoon. 1978. Priming in Item Recognition: Evidence for the Propositional Structure of Sentences. *Journal of Verbal Learning and Verbal Behavior* 17:403–417.

Reddy, Michael. 1979. The Conduit Metaphor. In A. Ortony, ed., *Metaphor and Thought,* 284–324. Cambridge: Cambridge University Press.

Rice, Mabel L., and John V. Bode. 1993. GAPS in the Verb Lexicons of Children with Specific Language Impairment. *First Language* 13, 113–131.

Rice, Sally. 1987a. Transitivity and the Lexicon. *Center for Research in Language Newsletter* 2.2. University of California, San Diego.

Rice, Sally. 1987b. Participants and Non-participants: Toward a Cognitive Model of Transitivity. Ph.D. diss., University of California, San Diego.

Rice, Sally. 1988. Unlikely Lexical Entries. *BLS* 14, 202–212.

Roberts, R. B., and I. P. Goldstein. 1977. *The FRL Manual.* Technical Report AIM-408. Cambridge, Mass.: MIT Artificial Intelligence Laboratory.

Rosch, Eleanor. 1973. Natural Categories. *Cognitive Psychology* 4:328–50.

Rosch, Eleanor, and Carolyn Mervis. 1975. Family Resemblances: Studies in the Internal Structure of Categories. *Cognitive Psychology* 7:573–605.

Rosch, Eleanor, Carolyn Mervis, Wayne Gray, David Johnson, and Penny Boyes-Braem. 1976. Basic Objects in Natural Categories. *Cognitive Psychology* 8: 382–439.

Rosen, Carol G. 1984. The Interface between Semantic Roles and Initial Grammatical Relations. In D. Perlmutter and C. G. Rosen, eds., *Studies in Relational Grammar,* vol. 2, 38–77. Chicago: University of Chicago Press.

Ross, John Robert. 1969. Auxiliaries as Main Verbs. *Journal of Linguistics* 1(1): 77–102.

Ross, John Robert. 1970. On Declarative Sentences. In R. A. Jacobs and P. S. Rosenbaum, eds., *Readings in English Transformational Grammar,* 222–272. Waltham, Mass.: Ginn.

Rumelhart, David E., and James L. McClelland, eds., 1986. *Parallel Distributed Processing: Explorations in the Microstructure of Cognition.* 2 vols. Cambridge, Mass.: MIT Press.

Sag, Ivan A. and Carl Pollard. 1991. An Integrated Theory of Complement Control. *Language* 67(1):63–113.

Salkoff, Morris. 1988. Analysis by Fusion. *Lingvisticae Investigationes* 12(1):49–84. Amsterdam: John Benjamins.

Sanches, M. 1978. On the Emergence of Multi-Element-Utterances in the Child's Japanese. Unpublished MS, University of Texas at Austin, Dept. of Anthropology.

Sapir, Edward. 1944. On Grading: A Study in Semantics. *Philosophy of Science* 2: 93–116.

Saussure, Ferdinand de. 1916. *Cours de linguistique générale.* Paris: Payet, 1973. Translated by W. Baskin. New York: McGraw Hill, 1976.

Schank, R. C., and R. P. Abelson. 1977. *Scripts, Plans, Goals and Understanding: An Inquiry into Human Knowledge Structures.* Hillsdale, N.J.: Lawrence Erlbaum.

Schieffelin, B. B. 1985. The Acquisition of Kaluli. In D. I. Slobin, ed., *The Crosslinguistic Study of Language Acquisition.* Vol. 1: *The Data,* 525–593. Hillsdale, N.J.: Lawrence Erlbaum Associates.

Schlesinger, I. M. 1971. Production of Utterances and Language Acquisition. In D. I. Slobin, ed., *The Ontogenesis of Grammar.* New York: Academic Press.

Searle, John R. 1983. *Intentionality: An Essay in the Philosophy of Mind.* Cambridge: Cambridge University Press.

Shibatani, Masayoshi. 1973. *A Linguistic Study of Causative Constructions.* Ph.D. diss., University of California, Berkeley.

Shibatani, Masayoshi, ed. 1976. *Syntax and Semantics 6: The Grammar of Causative Constructions.* New York: Academic Press.

Shieber, S. 1986. *An Introduction to Unification-Based Approaches to Grammar.* CSLI Lecture Notes no 4. Stanford, Cal.: Center for the Study of Language and Information, Stanford University.

Shieber, S., L. Karttunen, and F. Pereira, eds. 1984. *Notes from the Unification Underground: A Compilation of Papers on Unification-Based Grammar Formalisms.* SRI Technical Report 327. Menlo Park, Cal.: SRI International.

Simpson, J. 1983. Resultatives. In L. Levin, M. Rappaport, and A. Zaenen, eds., *Papers in Lexical-Functional Grammar,* 143–157. Bloomington: Indiana University Linguistics Club.

Slobin, Dan. 1970. Universals of Grammatical Development in Children. In W. J. M. Levelt and G. B. Flores d'Arcais, eds., *Advances in Psycholinguistic Research.* Amsterdam: North-Holland.

Slobin, Dan. 1985. Crosslinguistic Evidence for the Language-Making Capacity. In D. Slobin, ed., *A Crosslinguistic Study of Language Acquisition.* Vol. 2: *Theoretical Issues.* Hillsdale, N.J.: Lawrence Erlbaum.

Smolensky, Paul. 1986. Information Processing in Dynamical Systems: Foundations of Harmony Theory. In D. E. Rumelhart and J. L. McClelland, eds., *Parallel Distributed Processing.* Vol. 1: *Foundations,* 194–201.

Sproat, Richard W. 1985. On Deriving the Lexicon, Ph.D. diss., MIT.

Stowell, Timothy. 1981. *Origins of Phrase Structure.* Ph.D. diss., MIT.

Sweetser, Eve. 1990. *From Etymology to Pragmatics.* Cambridge: Cambridge University Press.

Talmy, Leonard. 1976. Semantic Causative Types. In Shibatani, M., ed., *Syntax and Semantics 6: The Grammar of Causative Constructions.* New York: Academic Press.

Talmy, Leonard. 1977. Rubber-Sheet Cognition in Language. *CLS* 13.

Talmy, Leonard. 1978. The Relation of Grammar to Cognition. In D. Waltz, ed., *Pro-*

ceedings of TINLAP-2 (Theoretical Issues in Natural Language Processing).
Champaign: Coordinated Science Laboratory, University of Illinois.

Talmy, Leonard. 1983. How Language Structures Space. In H. Pick and L. Acredolo, eds., *Spatial Orientation: Theory, Research, and Application.* New York: Plenum Press.

Talmy, Leonard. 1985a. Lexicalization Patterns: Semantic Structure in Lexical Forms. In T. Shopen, ed., *Language Typology and Syntactic Description,* vol. 3: *Grammatical Categories and the Lexicon,* 57–149. Cambridge: Cambridge University Press.

Talmy, Leonard. 1985b. Force Dynamics in Language and Thought. *CLS* 21.1, *Parasession on Causatives and Agentivity,* 293–337.

Tenny, Carol. 1987. *Grammaticalizing Aspect and Affectedness.* Ph.D. diss. MIT.

Thomason, R. H. 1992. NETL and Subsequent Path-Based Inheritance Theories. *Computers and Mathematics with Applications* 23(2–5): 179–204.

Touretzky, David. 1986. *The Mathematics of Inheritance Systems.* Los Altos, Cal.: Morgan Kaufmann.

Traugott, Elizabeth C. 1988. Pragmatic Strengthening and Grammaticalization. *BLS* 14, 406–416.

Traugott, Elizabeth C. 1989. On the Rise of Epistemic Meanings in English: An Example of Subjectification in Semantic Change. *Language* 65(1): 31–55.

Trechsel, Frank R. 1982. A Categorial Fragment of Quiche. Texas Linguistic Forum 20. Austin: Department of Linguistics, University of Texas.

Tuggy, David. 1988. Nahuatl Causative/Applicatives in Cognitive Grammar. In B. Rudzka-Ostyn, ed., *Topics in Cognitive Linguistics.* Philadelphia: John Benjamins.

Van Valin, Robert D., Jr. 1990a. Semantic Parameters of Split Intrasitivity. *Language* 66(2): 221–260.

Van Valin, Robert D., Jr. 1990b. The Linking Theory in RRG. Presentation at the Center for the Study of Language and Information, Stanford University.

Van Valin, Robert D., Jr. 1992. Incorporation in Universal Grammar: A Case Study in Theoretical Reductionism. *Journal of Linguistics* 28: 199–220.

Velásquez-Castillo, Maura. 1993. The Grammar of Inalienability: Possession and Noun Incorporation in Paraguayan Guaraní. Ph.D. diss., University of California, San Diego.

Visser, F. Th. 1963. *An Historical Syntax of the English Language.* Part 1, *Syntactical Units with One Verb.* Leiden: E. J. Brill.

Ward, Gregory, Richard Sproat, and Gail McKoon. 1991. A Pragmatic Analysis of So-Called Anaphoric Islands. *Language* 67(3): 439–474.

Wasow, Thomas. 1977. Transformations and the Lexicon. In P. W. Culicover, T. Wasow, and A. Akmajian, eds., *Formal Syntax,* 327–360. New York: Academic Press.

Wasow, Thomas. 1981. Comments on the Paper by Baker. In C. L. Baker and J. J. McCarthy, eds., *The Logical Problem of Language Acquisition,* 324–329. Cambridge, Mass.: MIT Press.

Watkins, R. V., M. L. Rice, and C. C. Moltz. 1993. Verb and Inflection Acquisition in Language-impaired and Normally Developing Preschoolers. *First Language.*

Wheeler, Daniel. 1970. Processes in Word Recognition. *Cognition Psychology* 1: 59–85.

Wierzbicka, Anna. 1986. The Semantics of the 'Internal Dative': A Rejoinder. *Quaderni di Semantica* 7:155–165.

Wierzbicka, Anna. 1988. *The Semantics of Grammar.* Amsterdam: John Benjamins.

Wilensky, Robert. 1982. Points: A Theory of the Structure of Stores in Memory. In W. Lehnert and M. Rengle, eds., *Strategies for Natural-Language Processing,* 345–374. Hillsdale, N.J.: Lawrence Erlbaum Associates.

Wilensky, Robert. 1986. Some Problems and Proposals for Knowledge Representation. Cognitive Science Report 40, University of California, Berkeley.

Wilensky, Robert. 1991. Extending the Lexicon by Exploiting Subregularities. Report UCB/CSD 91/618, Computer Science Division (EECS), University of California, Berkeley.

Williams, Edwin. 1983. Against Small Clauses. *Linguistic Inquiry* 14(2):287–308.

Wittgenstein, Ludwig. 1953. *Philosophical Investigations.* New York: Macmillan.

Zadrozny, Wlodek, and Alexis Manaster-Ramer. 1993. The Significance of Constructions. Unpublished manuscript, IBM T. J. Watson Research Center and Wayne State University.

Zaenen, Annie. 1991. Subcategorization and Pragmatics. Presentation at the Center for the Study of Language and Information, Stanford University.

Zaenen, Annie. 1993. Unaccusativity in Dutch: Integrating Syntax and Lexical Semantics. In J. Pustejovsky, ed., *Semantics and the Lexicon,* 129–161. Dordrecht: Kluwer.

Zaenen, Annie, and Adele E. Goldberg. 1993. A Review of Grimshaw's *Argument Structure. Language* 69(4):807–817.

Zubizarreta, Maria Luisa. 1987. *Levels of Representation in the Lexicon and Syntax.* Dordrecht: Foris.

Zwicky, Arnold. 1971. In a Manner of Speaking. *Linguistic Inquiry* 11(2):223–233.

Zwicky, Arnold. 1987. Constructions in Monostratal Syntax. *CLS* 14,

Zwicky, Arnold. 1989. What's Become of Derivations? Defaults and Invocations. *BLS* 15, 303–320.

Zwicky, Arnold. 1990. Syntactic Words and Morphological Words, Simple and Composite. *Yearbook of Morphology* 3, 201–216.

Index